Flexible Manufacturing

MANUFACTURING ENGINEERING AND MATERIALS PROCESSING

A Series of Reference Books and Textbooks

SERIES EDITORS

Geoffrey Boothroyd

Chairman, Department of Industrial and Manufacturing Engineering
University of Rhode Island
Kingston, Rhode Island

George E. Dieter

Dean, College of Engineering
University of Maryland
College Park, Maryland

1. Computers in Manufacturing, *U. Rembold, M. Seth and J. S. Weinstein*
2. Cold Rolling of Steel, *William L. Roberts*
3. Strengthening of Ceramics: Treatments, Tests and Design Applications, *Henry P. Kirchner*
4. Metal Forming: The Application of Limit Analysis, *Betzalel Avitzur*
5. Improving Productivity by Classification, Coding, and Data Base Standardization: The Key to Maximizing CAD/CAM and Group Technology, *William F. Hyde*
6. Automatic Assembly, *Geoffrey Boothroyd, Corrado Poli, and Laurence E. Murch*
7. Manufacturing Engineering Processes, *Leo Alting*
8. Modern Ceramic Engineering: Properties, Processing, and Use in Design, *David W. Richerson*
9. Interface Technology for Computer-Controlled Manufacturing Processes, *Ulrich Rembold, Karl Armbruster, and Wolfgang Ülzmann*
10. Hot Rolling of Steel, *William L. Roberts*
11. Adhesives in Manufacturing, *edited by Gerald L. Schneberger*
12. Understanding the Manufacturing Process: Key to Successful CAD/CAM Implementation, *Joseph Harrington, Jr.*
13. Industrial Materials Science and Engineering, *edited by Lawrence E. Murr*
14. Lubricants and Lubrication in Metalworking Operations, *Elliot S. Nachtman and Serope Kalpakjian*
15. Manufacturing Engineering: An Introduction to the Basic Functions, *John P. Tanner*
16. Computer-Integrated Manufacturing Technology and Systems, *Ulrich Rembold, Christian Blume, and Ruediger Dillmann*
17. Connections in Electronic Assemblies, *Anthony J. Bilotta*

18. Automation for Press Feed Operations: Applications and Economics, *Edward Walker*
19. Nontraditional Manufacturing Processes, *Gary F. Benedict*
20. Programmable Controllers for Factory Automation, *David G. Johnson*
21. Printed Circuit Assembly Manufacturing, *Fred W. Kear*
22. Manufacturing High Technology Handbook, *edited by Donatas Tijunelis and Keith E. McKee*
23. Factory Information Systems: Design and Implementation for CIM Management and Control, *John Gaylord*
24. Flat Processing of Steel, *William L. Roberts*
25. Soldering for Electronic Assemblies, *Leo P. Lambert*
26. Flexible Manufacturing Systems in Practice: Applications, Design, and Simulation, *Joseph Talavage and Roger G. Hannam*
27. Flexible Manufacturing: Benefits for the Low-Inventory Factory, *John E. Lenz*
28. Fundamentals of Machining and Machine Tools, Second Edition, *Geoffrey Boothroyd and Winston A. Knight*
29. Computer-Aided Manufacturing Process Planning, *James Nolen*

OTHER VOLUMES IN PREPARATION

Flexible Manufacturing

Benefits for the Low-Inventory Factory

John E. Lenz

CMS Research, Inc.
Oshkosh, Wisconsin

Marcel Dekker, Inc. New York and Basel

Library of Congress Cataloging-in-Publication Data

Lenz, J. E. (John E.)
Flexible manufacturing : benefits for the low-inventory factory/
John E. Lenz.
p. cm.
Bibliography: p.
Includes index.
ISBN 0-8247-7683-6
1. Flexible manufacturing systems. I. Title
TS155.6.L46 1988
658.5'1--dc19

88-39713
CIP

This book is printed on acid-free paper.

MARCEL DEKKER, INC.
270 Madison Avenue, New York, New York 10016

Current printing (last digit):
10 9 8 7 6 5 4 3 2 1

PRINTED IN THE UNITED STATES OF AMERICA

Preface

Flexible Manufacturing presents flexibility as a means for maintaining a desirable level of productivity. A production process having a higher degree of flexibility offers the advantages of a lower level of inventory, less balanced loading among stations or departments, and less need for short-term scheduling. The approach described in the book does not simply state the relationship among these variables but provides specific techniques for quantifying the give-and-take in meeting a production need.

The techniques discussed are derived from research of flexible manufacturing systems (FMS). After review, consultation, and design of more than one hundred FMS, one common characteristic was observed: Even if individual components are available 80% of the time, an FMS might achieve only 65% of its production capacity. This means that components will be available yet idle for 15% of the production time. Something within the FMS prevents component use for the entire time of component availability. The identification, formalization, and solution of this problem provided the basis for the research for this book.

The reduced productivity of the FMS is not due to deficiencies in the computer control system, computer reliability, software integrity, mechanization, or automation typically formed in an FMS. Rather, it is due to the production environment of low inventory and unbalanced characteristics. Flexibility was to substitute for these two production variables but was not able to fulfill its role. Even a real-time monitoring and scheduling system could not eliminate the inefficiencies found in FMS. Another explanation was needed; it is supplied by the manufacturing integration model.

Because this model of integration effects is not based on computer integration, automation, or even mechanization, it has a more widespread application in manufacturing, with FMS as only a subset of the model's potential applications. A colleague has stated: "If you can find a model that predicts performance of an FMS, then you have solved the difficult problem. The rest should be easy."

Just as my colleague has predicted, the application of flexibility to manual production systems, U-lines, assembly systems, and computer integrated systems, such as FMS, can be explained in terms of a uniform technique. This technique, referred to as integration effects, measures the amount of productivity lost under conditions of low inventory, unbalanced operation, and nonutopian flexibility.

Concerned with maintaining production capacity, an FMS supplier posed this question: "I know that these CNC machine tools will be available during the production period for 90% of the time. I can plan to lose 10% production capacity due to this effect. But how much additional capacity will I lose when I integrate these into an FMS?"

My initial response might have been to ask what he meant by reduced capacity. The FMS provides real-time tracking and scheduling software that should eliminate any inefficiencies due to failure of a single component. There should not be any additional loss in production and component efficiencies that would be consistent with that of stand-alone components. Flexibility and control systems should filter out any carry-over effects (integration effects) from one component to the next.

Having written this book, I would now respond by noting that the amount of additional capacity lost within the FMS will be equal to the percentage of time components that are available, but idle, during the production period. This characteristic is important only when it occurs on the bottleneck station; but if the bottleneck moves around, it is difficult to identify. The component will be idle for one of two reasons: shortage of material or blockage from a completed part. Neither utopian flexibility nor the ideal control system can prevent these situations from occurring in a low-inventory environment. The amount of production lost because of low inventory is a function of the degree of flexibility, balance, and work in process levels. A typical amount for most FMS is approximately 15%.

I do not mean to mislead the reader into thinking that this is a book about FMS. It is not. It is a book about flexible manufacturing, which involves using flexibility as a substitute for other production variables. FMS is the application of flexibility where it is implemented through real-time computer control, automated material handling, and work stations for a machine-limited process. This is only one of several possible applications of flexibility in the low-inventory factory.

Read and study this book carefully. It provides awareness and direction to many of the difficult challenges facing manufacturing today. Of course, I could not have completed this research without assistance and devotion from a unique club of individuals. I refer to this group as a club, but its membership is distributed worldwide, and seldom does one member share ideas or even know other members in the club. The club I refer to comprises manufacturing engineers, process engineers, and control engineers who have started a flexible manufacturing project and have remained through thick and thin (thick with problems and thin with money). Members of this club have not used an FMS project to provide a technological leap into higher management but have worked through the ups and downs to make flexible manufacturing a reality. Most of the members of this club realize that they are not "going places" within the organization. Instead they have found something that provides a great opportunity for the factory and have devoted time and effort over several years to make flexibility work. In the eyes of management, these club members are associated with projects that yield lower than expected rate of return. However, these club members have provided the success stories forged in a frontier without adequate information, guidance, and tools. For this reason, I have chosen to dedicate this book to the members of this club. I hope that after reading more of you will join the club.

Aside from the direct dealings with members of the flexibility club, I have benefited from the many others who have contributed to the completion of this book. These include my professors: Dr. George Hutchinson of the University of Wisconsin—Milwaukee and Dr. Joseph Talavage of Purdue University; Dr. Talavage assisted me in developing my first model of flexible manufacturing, and Dr. Hutchinson helped me to pursue a scientific method during the development of the manufacturing integration theory. Also, I have appreciated the assistance of Cyndie, Sharon, Chris, Jan, Jim, Ilkka, Parmjit, Henrik, Aksel, Jens, Calle, Moshe, Heidi, and the employees of CMS Research. Special appreciation goes to my wife, Kathleen, and my children, John and Annelise, for their continued support of this ambitious project.

John E. Lenz

Contents

Flexible Manufacturing

1

Introduction to Automation

The McGraw-Hill Encyclopedia of Science and Technology provides the following definition for automation: "A coined word having no precise, generally accepted technical meaning, but widely used to imply the concept of development or use of highly automated machinery or control systems." With this definition, automation can be all things to all people. And it is! There are few other subjects which lack the paradigm needed for research and logical development. Because of this lack of organization, the term "island of automation" (Fig. 1-1) has become an accurate representation of automation in today's factory. Why is the factory automated to various degrees where some processes are performed completely with machines and yet others remain as manual labor intensive tasks? Where has the integration gone?

1.1 COINING THE TERM "AUTOMATION"

Henry Ford envisioned the mass production of the automobile. With this, he initiated many new concepts, the most important being the transfer line. From 1900, Ford continued to incorporate new manufacturing technologies into the mass production of the automobile. In 1946, D. S. Harder, a manager at Ford, coined the term "automation." He used it to describe machinery being developed for the mass production of the engine block. By today's standards, Mr. Harder would be referring to an integrated material handling system used to synchronize several independent operations. From this,

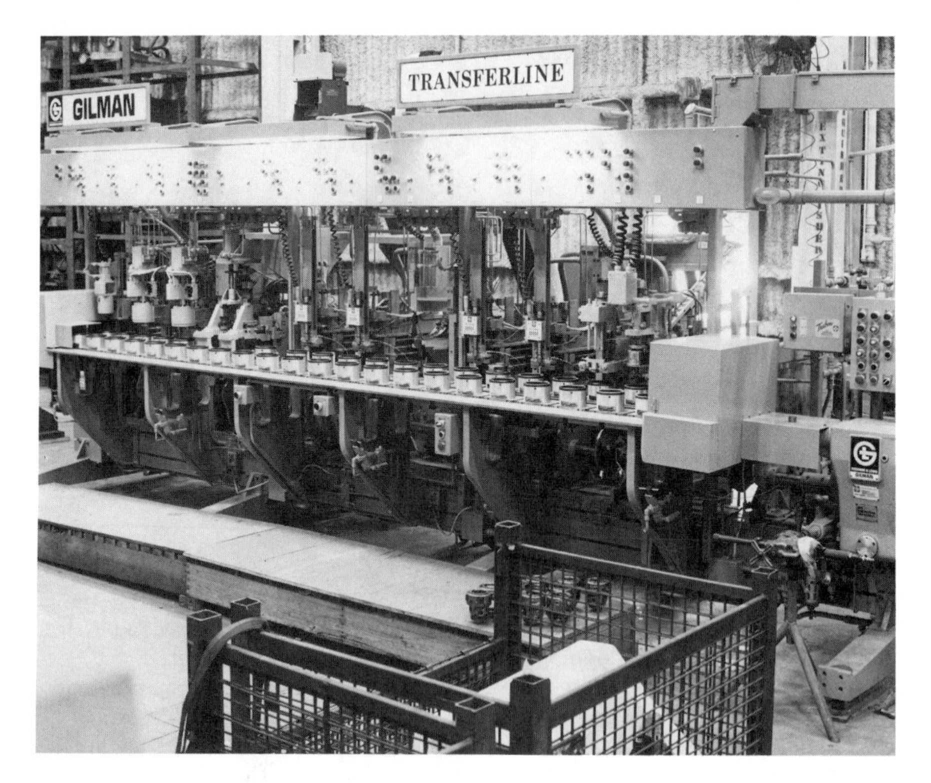

FIGURE 1-1 Island of automation.

Ford formed the first automation department in 1947. This department was assigned the objective of finding suitable processes which could be "automated." Most of these projects resulted in advanced application of material handling to integrate various operations. This group provided more than pure mechanization, as they reviewed the process for automation suitability.

This started a new approach to manufacturing, one where new technologies and innovative ideas would become commonplace. In fact, the generation of ideas was not limited to engineers. Automation became a popular subject with social scientists as well. It is from these new technical ideas that automation obtained its "ominous" tone.

1.2 AGE OF AUTOMATION

The term "automation" first appeared in print in the October 21, 1948, issue of *American Machinist**. In this article, only the technical features of automation were presented but they focused on the control characteristics of the hardware (i.e., decision making). This prompted many social scientists to report on the future impact of this new technology. Figure 1-2 summarizes the major events in the age of automation.

In 1950, Norbert Weiner pulished a book entitled *The Human Use of Human Beings*. In this book, he predicted that the use of automatic control would make the automatic factory a common reality in 25 years. This would be possible from the result of computers being applied to production machinery. He concluded his ideas by stating that these changes in our method of production would bring a severe depression to the economy. This started the ominous tone of automation.

Date	Event	Future social impact
1948	Coined term automation	New technology
1950	Norbert Weiner publication	Automated control will be a reality within 25 years
1952	John Deibold publication	Feedback control systems
1962	Published reports recognize automation	Society threatened with expanded automation
1963	Congressional investigations	Automation leads to unemployment
1966	Trade publications	100% total unemployment by 1980?
Today	Computerized adaptive controls	"Island" type applications

FIGURE 1-2 Island of automation.

*There is some discrepancy as to whether this report was the first. *Life* magazine was to coin the word in an article in 1947.

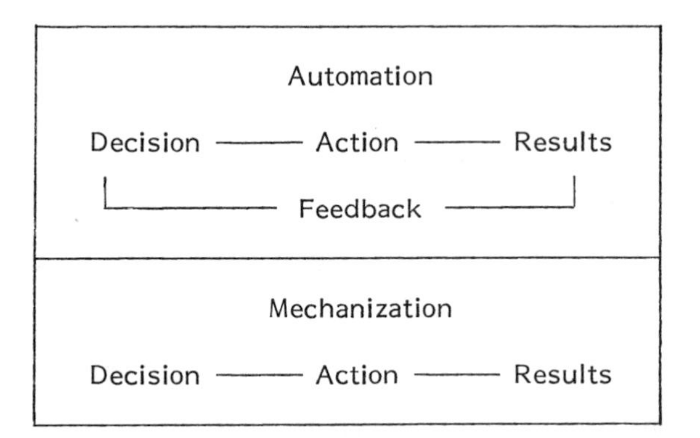

FIGURE 1-3 Automation/mechanization.

Probably the most widely read and referenced printed material on automation was written by John Diebold in 1952, entitled *Automation*. In this book, he focused on the role of control systems and their similarity to the decision making process. He also associated the process of learning as being comprised of iterative "feedback" to correct or reinforce decisions. Diebold described this learning or use of feedback as an integral component of automation. This provided the means to distinguish mechanization from automation (Fig. 1-3).

Mechanization represents the direct replacement of humans with machinery. Often in these replacements, the process is unaffected and benefits are determined by reduction of labor cost or cost due to hazardous environments. In these applications, the machinery is unable to adapt to changes in its environment. For example, if a part's orientation is slightly different from what is expected, a robot might not be able to latch onto the part. An interruption in service will occur (a fault) and operator intervention will be needed to restore service (fault recovery).

Automation includes the same machinery found in mechanization, but the control for the machinery is able to adapt to *known* and *limited* changes in the environment. This is done through the use of feedback control systems, outlined in Diebold's book.

From this distinction of automation and mechanization (Fig. 1-3) it is clear why mechanization did not reach the level of threat as did automation. Automation was interpreted to mean that machinery would replace labor and computers would replace decision making. This was seen as a threat to society by many social scientists and a threat to the jobs of many workers.

Leaders of American labor became outspoken and were requested to provide representation to the investigations of the impact of automation. One such expression was provided by Walter P. Reuther, president of the United Automobile Workers, when he appeared before the House Subcommittee on Economic Stabilization in October 1955:

> A rough casting comes out of a foundry. The foundry is automated. . . . The rough casting comes into the machining operation. It is fed into the machine, the first operation is to machine the top of the cylinder block and the bottom where the crankcase goes on. It takes 13 seconds to do that operation. It goes "whoosh" and it is done. The rest of the operations are worked from those two machined surfaces. The automatic lathe then bores the cylinder block. After the cylinder block is bored, the electric eye measues the block and, if it is not the exact size required, an electric impulse goes to the brain of the machine and the tool is adjusted, a new cut is taken, it comes back and the electric eye measures it. . . . That machined block comes out the other end in 14.6 minutes without a human hand touching it.
>
> Some years back, we made the first engine block in 24 hours, from a rough casting to the finished block. To machine a rough casting to a finished motor block in 24 hours was hailed an unprecedented technological achievement. Then we got it down to 9 hours. . . . Now we've jumped to 14.6 minutes.

In 1962, an organization in Michigan published reports of an unmanned bakery in Chicago. It claimed that this bakery was producing hundreds of loaves of bread each day with only a handful of workers. Another organization wrote of the ghost towns of Texas, where the oil refineries had become so automated that whole towns had become vacant. All of these threats to society reached their peak in 1963 when President Johnson initiated the elimination of the triple threats to society. The three threats were war, poverty and automation. In 1963, social scientists reported at a heaving of the Congressional Subcommittee on the three threats of society that ". . . automation will result in total unemployment within 20 years."

This statement, which was made to the Congress of the United States, was taken very seriously. Social scientists wrote more and more on how the United States would be totally unemployed by 1980.

But in none of these reports did they describe specifically the details by which this change would take place. They only could envision the end result which, in simple replacement, would appear to be extremely cost-effective.

1.3 EVOLUTION OF MANUFACTURING

But as we now know, the evolution to the automation factory has been risky, expensive and laiden with failures, with a scattering of success stories. These success stories are referred to today as the "islands of automation."

The primary reason the automatic factory is not a reality today is that the computerization of human decision making is an extremely difficult, misunderstood and costly process. The design and invention of machinery to perform human actions is not a difficult task in itself. The difficulty comes in providing machinery with the capability to adapt to changes in its environment. Today, adaptive controls have been developed for known and planned for changes, but until this machinery can adapt to unexpected changes (as humans do so well), the automation factory will remain a dream.

Today's research focuses on providing machinery with sensors able to detect changes in the environment. With this information, computer programs can direct the machinery in a seemingly responsive operation. Today, problems are the accuracy, timeliness, cost and effectiveness of these sensors as applied in a variety of environments.

The technology that has shown the efficiency or weaknesses of automation control is the Flexible Manufacturing System (FMS). But before the specific characteristics of flexible manufacturing are presented, a general model of manufacturing is presented in Chapter 2. This model provides a basis for identifying the benefits of flexible automation over alternatives. In these terms, flexible manufacturing provides an alternative to inventory or balanced operations. However, flexibility often implies computerized control, the most difficult hurdle to overcome in automation. Now, technologies are evolving to assist in this problem. One of these new technologies is Artificial Intelligence.

1.4 ARTIFICIAL INTELLIGENCE

An emerging technology which addresses the weakness in adaptive process control is Artificial Intelligence (AI). This technology is based on the intelligent, expert use of information provided by sensors. Sensors and computer programs can be adapted to respond

1. Rule based theory
   Example: A doctor's diagnosis
2. Theory based
   Example: Temperature control theory

FIGURE 1-4 Knowledge representation.

to known or predictable situations which might arise in the factory environment. But when a new situation arises which has not been planned for, the automation immediately takes on mechanization characteristics. Therefore, automation depends on the experience and foresight of those who develop it.

Artificial Intelligence provides an ability for the control to respond to situations which were not known at the conception of the automation. Artificial Intelligence control software will permit recognition of a new situation and deduction of a response from some "thought" process. This process permits what appears as "reasoning by the computer." However, this reasoning capability will still be limited to the foresight and experience of the computer programmer, not the manufacturing engineer. The limits of automation control will still exist, but they will arise from different sources. A summary of knowledge representation is shown in Fig. 1-4.

1.5 CONCLUSION

Automation means different things to many people. However, it is important to clarify the distinction between mechanization and automation. Mechanization does not require controls which can adapt to a changing environment and provide benefits in the form of "islands." Automation cannot be beneficial until it provides sensors which can detect changes in the environment and report them accurately, computer software which can diagnose the situation and reach the proper response and mechanics which can be instructed to carry out these responses.

Chapter 2 contains a description of a general model for the study of integrated manufacturing. This model provides a framework for identifying the role of flexibility and its benefits for the automated factory.

2

General Model of Manufacturing

All problems need a model on which formal methods of study can be based. Some disciplines, such as organizational behavior, rely exclusively on models, whereas other disciplines rely on proven mathematical theories. In either case, these rules form a foundation from which techniques can be compared, measured for effectiveness and from which further knowledge can be obtained.

Because of the need for such a manufacturing model, a host of new techniques have emerged. It is necessary to relate and measure the suitability and effectiveness of these techniques, one of which is flexible manufacturing. To adequately present the benefits and costs of flexibility, they must be compared to manufacturing alternatives. Before this comparison can be made, the alternatives must be identified. Identification of flexibility and its alternative to manufacturing are the subjects of this chapter.

Another important concept which is presented in this chapter is the establishment of the basic principle that manufacturing strategy is purely a choice of alternatives (Fig. 2-1). This implies that the absence of one means the presence of another. For example, many researchers and consultants have described the benefits of low inventory production environments. Granted, there are obvious benefits of low inventory, but whenever something is taken away from the manufacturing process, something must be put back in its place (Fig. 2-2). Increases in flexibility will permit the opportunity to give up something else. Determining which alternative can be given up for flexibility is the primary role of the Manufacturing Integration Model (MIM).

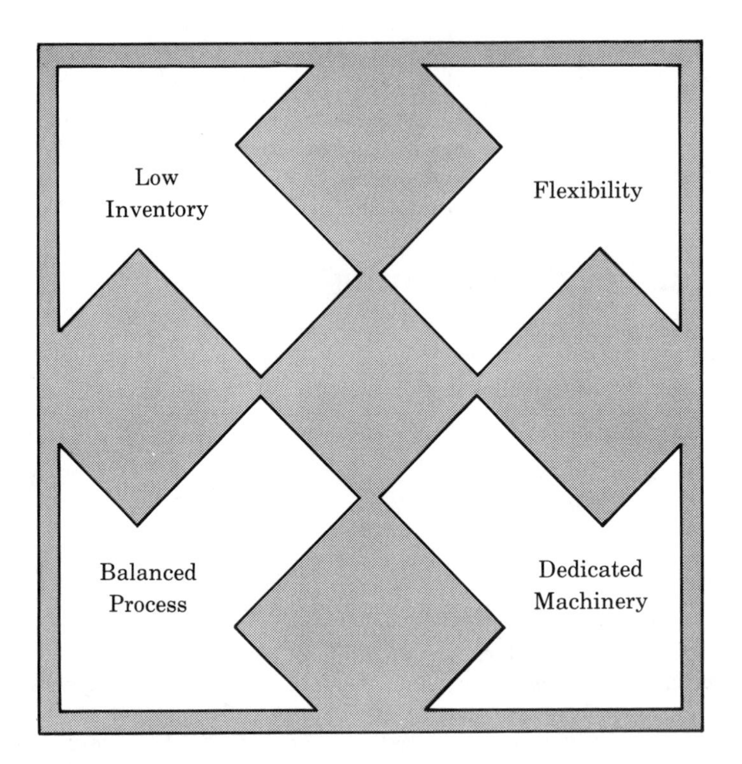

FIGURE 2-1 Alternative strategies.

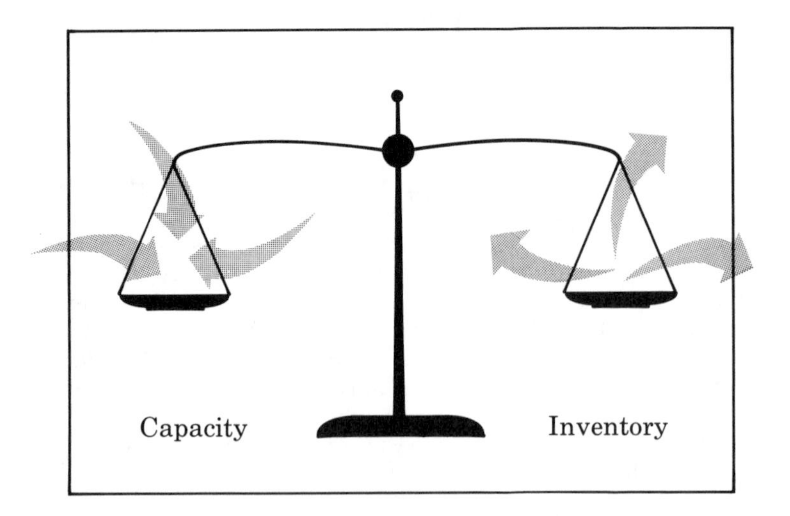

FIGURE 2-2 Balance of manufacturing.

2.1 MANUFACTURING INTEGRATION MODEL (MIM)

Why focus on integration of manufacturing rather than on other parameters, such as production or quality? This is best answered by stating manufacturing's overall goal to produce a product in the *most cost-effective* way. This can be interpreted to mean that the overall objective (Fig. 2-3) is to purchase a minimum amount of capacity (capital investment) and utilize it in the most effective way. Further, this requires obtaining production capacity which will be adjusted by interruptions in service due to equipment failures, operator breaks, lack of material or the inability to deliver material because of blockage. All of these are easily understood in terms of their direct effect on the specific work station. But it is equally important to identify the effects of lost capacity upon one station and its carry-over effects on other work stations. These carry-over effects are defined as Integration effects (Fig. 2-4).

Integration effects provide a means to measure the adjustments to production capacity for the entire manufacturing facility. The need to study these integration effects did not exist in 1975 because there were means to "filter out" these carry-over effects with high inventory levels or balanced operation times. However, the invention of the flexible manufacturing strategy has established the need to identify, understand and measure these effects, which are most prominent in the low-inventory, unbalanced operation environment. Chapter 3 discusses why this is so.

If there is anything different about flexible manufacturing from other strategies of manufacturing, it is the unpredictable nature of

★ Minimize capacity (capital investment)

★ Maximize utilization (productivity)

FIGURE 2-3 Manufacturing objective.

★ Station is idle because of lack of material

★ Station blocked: Unable to deliver material

FIGURE 2-4 Definition of integration effects.

★ Low inventory production

★ Unbalanced loads of work stations

FIGURE 2-5 FMS characteristics.

its production rates. This has prompted a need to model and explain the unique characteristics of flexible manufacturing. I wish I had recognized flexible manufacturing in 1975 as a strategy that had integration effects which could not be filtered out. I wish I had then proposed the need to study the lost capacity due to integration. It has taken me eight years and hundreds of projects to finally recognize and formulate a model of these effects. However, one benefit is that these integration effects are not unique to the flexible manufacturing system (FMS). The FMS was the first technique where these effects showed up; however, any low inventory, unbalanced production environment will experience the same adjustments to capacity, as observed in the FMS (Fig. 2-5). The degree of automation has little to do with the effects of integration on production capacity (see Chapter 3).

In the presentation of the Manufacturing Integration Model (MIM), the variable which is to be measured is "net production capacity." "Net" value is determined from "gross" capacity which has been adjusted due to integration effects. This definition was derived from a question posed during the design of a flexible manufacturing system:

> The designer of an FMS was including several machines which had been installed as individual work stations before. Over several installations, he knew that these machines could be expected to be available 90% of the time. However, he wanted to know how much "additional" capacity would be lost because these machines would be operating in a closed system with limited inventories. What he was asking was what additional capacity would be lost due to integration effects.

This has become the focus of MIM.

2.2 MANUFACTURING INTEGRATION MODEL DEFINITION

A basic MIM is shown in Fig. 2-6. These three equations provide a means to explain the real or net production of a manufacturing

Net production = gross production − station unavailability
− integration effects

Gross production = Σ (part volume * operation durations)

Station unavailability = F [break down, operator/tool availability, set-up times, rejected parts]

Integration effects = F [inventory level, balanced loadings, flexibility]

FIGURE 2-6 MIM model.

facility. The first equation shows the relationship between gross production and net production. Gross production is defined as the total production capacity of the manufacturing facility. It can be computed from use of production hours which are available, part operation durations and part process plans. The net production is the actual or real production which can be observed from the operation of the manufacturing facility. The difference between the net and gross production is represented as interruptions in service created throughout the facility.

The lost capacity due to station unavailability is the measure of the production lost directly due to a station unable to perform an operation. These interruptions include a breakdown, the unavailability of an operator or tools and set up time for different operations. Rejected parts also account for station unavailability. Figure 2-7 shows these variables.

	Unit of measure
Breakdown ⟶	production lost during repair time
Operator availability ⟶	production lost due to operator availability
Tool availability ⟶	production lost due to tool availability
Set-up time ⟶	production lost due to set-up time
Rejected parts ⟶	production lost due to inferior quality

FIGURE 2-7 Definition of station availability.

Gross production period:	1 shift = 8 hours
Operator breaks:	−45 minutes
Machine maintenance:	−30 minutes
Set-up time:	−15 minutes
Rejected parts:	−15 minutes
Net production period:	$\frac{\text{6 hours 15 minutes}}{\text{8 hours}}$
Efficiency factor:	$\frac{6:15}{8:00} = 78\%$

FIGURE 2-8 Planning for station unavailability.

The effects of processing a rejected part are similar to the effects caused by interruption in station service. When a station fails, it stops supplying parts to subsequent operations. The same is true of a rejected part. The time spent processing this rejected part is a direct adjustment to station availability because this process did not supply a part to subsequent operations. Thus, the time spent processing a rejected part will have similar integration effects as if the station had failed for the same period.

The difference between station failure and processing a rejected part is in the rework which results. This need for rework can be viewed as the "carry-over" effect or as a window of inventory effect to stations where the rework is performed. The processing of a bad part directly affects the capacity of the work station, and its integration effects are the creation of reworks. This rework can be viewed as having the same effects as a window of inventory upon the production capacity.

The method for planning for station unavailability is a well established procedure. Even though station unavailability can appear in a variety of forms, its effect is accounted for by use of a single efficiency factor. This efficiency factor (Fig. 2-8) is used as a percentage adjustment to the production planning period and accounts for all lost capacity on the individual work station.

In this manner, the specific amount of lost capacity due to rejected parts or station breakdown and repair is not identified.

Gross production	50 parts/hour
Efficiency factor	80%
Production lost to station unavailability (50 * 0.2)	10 parts/hour

FIGURE 2-9 Station availability and lost production.

Instead, a global percentage, determined from the experience of manufacturing engineers, is applied. For example, transfer line type manufacturing systems might use an efficiency factor of 70% whereas job shop environments might use 80% for planning.

This factor provides a means to adjust gross production capacity to account for breakdowns, operator/tool availability, set up times and rejected parts (Fig. 2-9). In this method, net or real production can be predicted with a high degree of accuracy, due to the ability to accurately predict an efficiency factor. However, this method assumes that there will be no additional lost capacity due to integration, for instance when a station sits idle waiting for a part because an upstream station has interrupted the supply of parts.

For example, suppose an efficiency factor of 80% is used. Planning is then based upon a station being busy 80% of the time and idle 20% of the time. At this point, there is no need for estimating station down time because it is accounted for in the efficiency factor (Fig. 2-10a).

But as the manufacturing facility starts to operate, the work station will experience interruption in their operation. The desired environment is one where planned idle time is equal to actual down time (Fig. 2-10b). In this environment of swapping idle time for down time, net production can be accurately predicted. But when integration effects are present, there is not a simple replacement of idle time for down time.

Instead, stations will become idle because of a lack of material or because they are blocked and unable to deliver their material. Thus, the down time will create carry-over effects which will reduce the actual busy percentage to below the planned level (Fig. 2-10c). How much busy time is lost to down time due to carry-over effects (Fig. 2-10d) is the focus of the next section.

The third variable for determining net production is the production capacity lost due to the carry-over or integration effects from interruptions in station availability. These integration effects are

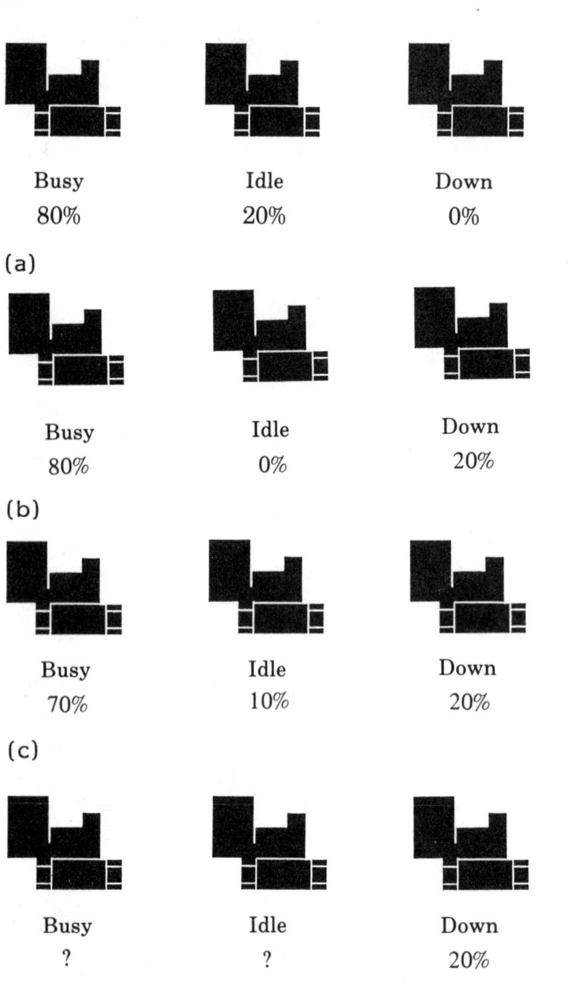

FIGURE 2-10 Planned operation with and without integration effects.

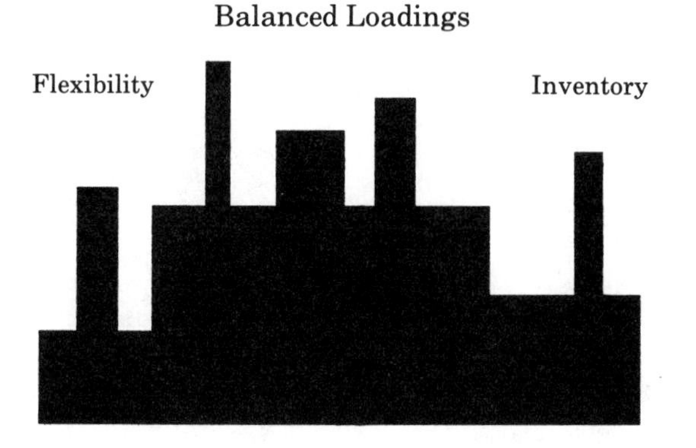

FIGURE 2-11 Facility productivity factors.

defined as a function of three variables. Consequently, the amount of lost capacity due to integration is determined by the inventory level, balanced loadings and flexibility present in the manufacturing facility (Fig. 2-11). The inventory level represents the total number of parts which have accumulated between work stations. The balanced loads will represent the degree of similarity in operation durations or, more generally, the balance of theoretical requirements for the various work stations. This can be thought of as a general case of line balancing for the transfer line. Finally, flexibility represents the number of alternative stations where a specific operation can be performed. One means to quantify this flexibility would be to think in terms of the number of different *paths* a part may take through the production facility (see Chapter 10). The degree of flexibility is increased with identical stations which are capable of performing the same operation and also with equipment or people, potential backups to operations. The relationship between these three variables and integration effects is described below.

2.3 INVENTORY'S RELATION TO INTEGRATION EFFECTS

Inventory is the number of parts which are permitted to accumulate between work stations. When the total operation time for the inventory level which has accumulated for a work station is less than the repair time of a previous work station which has failed, this station will be idled because of lack of material. This is an integration effect. To eliminate this effect, inventory levels must increase to exceed the longest possible repair time of a preceding station. In fact,

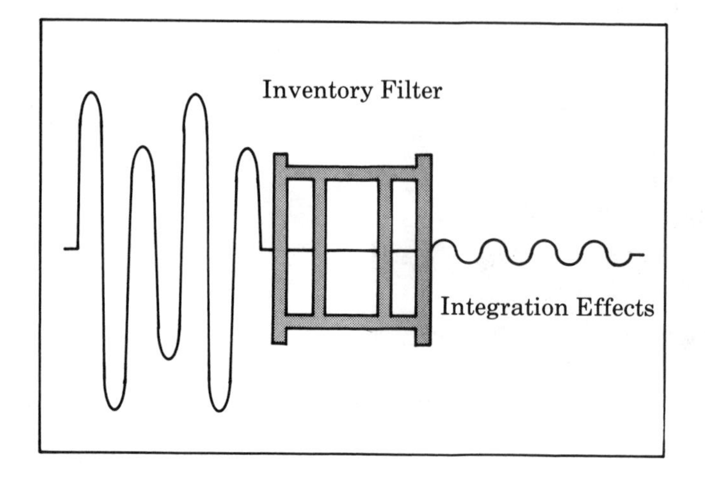

FIGURE 2-12 Filtering out effects.

inventory can be raised until it completely filters out any integration effects (Fig. 2-12). Because there are no carry-over effects being passed from station to station, the actual production can be computed directly from the station availability. In this case, the realistic production capacity is equal to the gross capacity that need only be adjusted for individual station availability. This describes the traditional job shop environment.

Suppose a very simple production system exists, where each part must travel through three consecutive stations (Fig. 2-13).

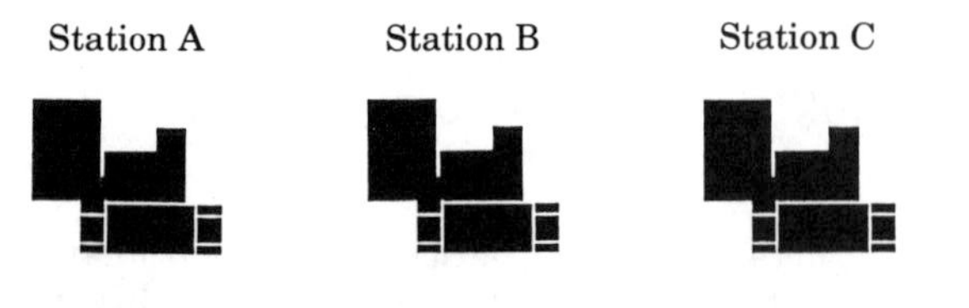

Station A: Brokendown, unavailable for 30 minutes
Station B: Unavailable for 10 minutes due to inventory
Station C: Unavailable for 5 minutes due to inventory

FIGURE 2-13 Stations A, B and C.

Now, suppose station A fails for 30 minutes. If there is more than 30 minutes of work accumulated for station B, station B's production rate will not be affected. But if there is less than 30 minutes of work (say 20 minutes), station B will be idle for ten minutes without any work. Suppose that there are five minutes of inventory accumulated between B and C. The ten minutes of lost capacity on B will become a five minute capacity loss in C as well. So, a 30 minute interruption in the service of station A has integrated effects of 10 minutes on station B and five minutes on station C. One way to visualize this effect is to think of this lost capacity as due to a "window of inventory." Also, note that in a production system which maintains consistently low inventories throughout the facility, these windows will move through the facility like an air bubble moves through a water pipe. Thus, interruptions can occur in an initial process and, eventually, a final process can be idled for lack of material.

Just-In-Time (JIT) techniques have become prominent in many production facilities. However, if these are implemented with the sole purpose of reducing inventories, the amount of lost capacity due to integration effects will increase. Stations will be found empty without any work. In order to reduce these integration effects (close-up windows of inventory), adjustments to the other three variables of MIM will be necessary.

In application, many manufacturing firms implement JIT techniques through material tracking procedures which lower inventory (Fig. 2-14). They react by purchasing additional capacity to make these levels more desirable. But in effect, what has happened has been a trade-off of inventory for additional station availability. Rather than have this take place as an action—reaction (Fig. 2-15), MIM provides a tool which can evaluate alternative trade-offs for cost-effectiveness. The purpose of MIM is not to identify the weaknesses of any technology, but to provide a means to understand and measure the *give and take* relations in manufacturing strategies.

In summary, inventory levels can be raised to completely eliminate (filter out) integration effects. Windows of inventory will never appear and each station can operate as an island in the

Take: Remove work in process inventory
Give: Simple shop floor material tracking procedure

FIGURE 2-14 JIT: Give and take.

Action:	Lower inventories; integration effects increase because control has limited information
Reaction:	Increased gross capacity to offset lost capacity due to integration

FIGURE 2-15 JIT: Action and reaction.

production process. The amount of inventory needed for this filtering out is determined from the repair time of previous stations. If repair times are short, only a few parts might be needed, whereas if repair times are long, several parts will be needed to prevent a window of inventory from occurring.

Just-In-Time techniques provide a means to lower inventories which will result in increased integration effects. Stations will be empty because of a lack of material. What needs to be studied are the benefits of lower inventory versus the cost of reduced production capacity or the cost of acquiring additional capacity. Some techniques indicate that windows of inventory can be avoided with sophisticated scheduling. But no matter how sophisticated the control techniques are, they can never guarantee that a station will be kept busy whenever it is available. Sophisticated scheduling can smooth out some integration effects, but the amount of lost capacity will depend on the inventory level, balance loadings and flexibility. Scheduling is most effective in the high inventory production environment. But here, integration effects are filtered out by inventories and the problem becomes making the right thing at the right time. Chapter 3 contains a detailed explanation of scheduling's role in manufacturing.

2.4 BALANCE LOADING'S RELATION TO INTEGRATION EFFECTS

To understand what is meant by balancing, it is best to start with a special case and then to generalize. Traditionally, balancing is the similarity between operation duration along a transfer line. For example, suppose a part must travel through two stations, A and B (Fig. 2-16). If the operation time on A is one minute and the operation time on B is 30 seconds, station B can never be utilized more than 50% of the time. This means that station B will be idle, waiting for material 50% of the time. To eliminate this integration

Station A

One Minute Cycle

Station B

Thirty Second Cycle

FIGURE 2-16 Station A and B cycles.

effect, the operation duration on A must be equal (or balanced) to the operation duration on B. A line of perfect balance means no lost capacity due to integration effects (Fig. 2-17). However, to utilize this concept in any type of production facility, it needs to be generalized. In the general form, the balance can be computed from the total requirements for a particular operation or station group. These station loads are the minimum amount of use to meet production requirements. For example, if a production plan requires 100 total hours of milling and 70 total hours of drilling and there are an equal amount of milling and drilling hours available, the drills will be less utilized than the mills or integration effects will exist. One simple means of eliminating this effect is for some of the drill stations to perform some milling operations. This will balance the loadings by use of flexible equipment. Again, the give in this production process is a need for perfect balance and the trade-off is stations with flexible tooling. The cost and benefits must be studied to find the best mix of give and take that yields an effective production system.

Some researchers of FMS propose that a high degree of flexibility is more effective when production systems are unbalanced. One researcher stated in 1977 that FMS should not be balanced. It has taken me nine years to understand his meaning: that

Station A

One Minute Cycle

Station B

One Minute Cycle

FIGURE 2-17 Balanced operations of stations A and B.

Give:	Ensure bottleneck station never experiences integration effects
Take:	Increased production capacity due to lost production from integration effects

FIGURE 2-18 OPT: Give and take.

perfectly balanced systems will result in several potential bottleneck stations. A bottleneck is any station with an availability that directly affects production rates. With perfect balance, the bottleneck station will move from station to station, determined by the frequency of breakdowns and repair durations. With so many potential bottlenecks in the production system, only a system with utopian flexibility consisting of identical stations could provide appropriate adjustments to reduce the integration effects. One technique which addresses the problem of bottleneck stations is the Optimized Production Technique (OPT).

The Optimized Production Technique theory proposes that integration effects can be minimized by keeping the bottleneck station from ever becoming idle. It states that one should identify the bottleneck and prevent it from experiencing any integration effects (Fig. 2-18). The practical application of this theory is best suited for those production systems which have a clear identifiable bottleneck station. This is most likely to occur in unbalanced systems. However, as systems become more balanced, the bottleneck will move around as often as a station fails or is repaired. In this environment, OPT will require timely, accurate data of the current production status. From this data, it can identify and direct attention to the "current" bottleneck. But if the current status of all stations and operation completions are not maintained, OPT will be using data which does not reflect the actual production status. In this case, it will not be able to identify the bottleneck and thus will not be able to manage integration effects. The critical nature of the need for timely, accurate data is best suited to computerized manufacturing. But such expectation of manual production process will reduce the effectiveness of OPT. Integration effects can be managed with sophisticated control (Fig. 2-19), but it requires perfect information, an unrealistic condition for most production environments.

OPT does filter out integration effects but is most effective where clear bottlenecks exist. Its effectiveness is directly related to the availability of timely, accurate data, and in the case of a clear

Action:	Install control system which readily identifies bottleneck station; integration effects increase because control requires timely information
Reaction:	Unbalance the process so that a clear bottleneck exists and does not move around

FIGURE 2-19 OPT: Action and reaction.

bottleneck, less data is needed. Thus, OPT is most effective where unbalanced loads exist.

In summary, balancing can reduce integration effects. In fact, the perfectly balanced transfer line has no integration effects other than those due to interruptions in station availability. But the carry-over effects can be filtered out. The definition of balance generally means minimum requirements for production functions. As these become unbalanced, single bottlenecks can occur in the production system. Scheduling techniques for OPT have been developed which attempt to minimize the lost capacity due to integration by prompt identification of bottlenecks. However, as with any scheduling techniques, they require timely, accurate information to be effective.

2.5 FLEXIBILITY'S RELATION TO INTEGRATION EFFECTS

Flexibility means many things to many people. However, all have a common characteristic of offering alternative paths for a part to take through a production facility. A path is defined as a unique sequence of work stations which a part uses to complete its required operations. The more paths that exist (current or potential), the greater the flexibility of the production facility.

Describing the relationship between flexibility and integration effects requires an explanation of flow time and its impact upon production. One primary characteristic of flexibility is its impact upon flow time. Flow time is the throughput time of a part through the production facility. As the part has more alternative paths (a higher degree of flexibility), its flow time will deviate less about some average time. It is important to note that higher levels of flexibility might reduce flow time, but its primary characteristic is

Flexibility ——> ——> measured with change in flow time

FIGURE 2-20 Units of measure for flexibility.

consistent flow times or smaller variations about some average. Automation reduces flow time, and flexibile automation, such as flexible manufacturing, provides reduced flow times which have small deviations (Fig. 2-20).

Thus, the relation of flexibility to integration effects is described in terms of flow time relations to integration effects. When a part stops moving (such as in any storage device), its flow time will increase. If this stopping is sporadic, its flow times will have larger deviations from one part to the next. This stopping of part flow is another means of describing a window of inventory. Consider the production facility as a flow process of parts moving through work stations. As stations fail, these flows are interrupted and parts spend additional time in storage. This interruption in the flow will create a window of inventory. But as the number of alternative paths increases, part flow is less likely to be interrupted and smaller windows of inventory will result.

In terms of integration effects, flexibility is being added in exchange for low inventory production. But flexibility is often associated with a high degree of automation. Complex computer control systems are necessary to properly implement flexibility. Granted, it is much easier to take advantage of flexibility with an automated material handling system and an on-line information system, but it is not necessary.

There are two methods for obtaining flexibility: computers and people control (Fig. 2-21). In fact, some Japanese techniques utilize flexibility by use of workers. For example, operators are assigned to specific stations in the production facility. However, if one operator starts to fall behind, he signals through a beacon light

1. Computer control ——> machine limited process
2. People control ——> labor limited process

FIGURE 2-21 Types of flexibility.

Give:	Elaborate alternative paths for parts through a process
Take:	Shorter and more consistent flow time

FIGURE 2-22 FMS: Give and take.

that he is having problems keeping up. Operators who work at downstream operations then stop their current assignment and reassign themselves to the problem area. In a sense, they provide alternative paths for parts by providing a variable number of workers at stations. But they only increase this flexibility as it is needed. As a result, flexibility in the work force provides consistent flow times in the production facility. Of course, such flexibility is only possible when the process lends itself to labor input. The design of work stations and operations must have characteristics which lend themselves to a flexible labor force.

No matter how flexibility is implemented, the give and take upon the production facility results in a more predictable flow time (Fig. 2-22). This predictable flow time provides consistent flow of parts through the facility which will keep windows of inventory to a minimum. In fact, if a production facility has utopian flexibility, windows of inventory and integration effects can be filtered out completely. For example, consider an FMS with three identical machines. Suppose parts travel to any one of these three machines and then return for unloading. If one station fails, its interruption in service will have no integrated effect upon the others because they are identical replacements of one another and parts do not travel from one to another. In this case, the actual production rate of this FMS will be equal to the available capacity and there will be no losses for stations sitting idle, waiting for work. However, if parts must travel from one station to another and some are dedicated for specific operations, integration effects will occur. This might explain why flexible manufacturing has tended towards production cells where homogeneous operations are performed.

In the summary seen in Fig. 2-23, the higher the level of flexibility, the more consistent the flow times for parts. The production facility will be able to maintain a consistent velocity of part movement. If part flows are consistent through high degrees of flexibility, integration effects can be filtered out. As stations or workers become more dedicated to operations and parts must flow through their specialized stations, integration effects will increase, which will decrease the production capacity of the facility.

Action:	Install FMS with total automation; integration effects increase because control system has less than perfect information
Reaction:	Additional people for maintenance, lower production, longer flow times

FIGURE 2-23 FMS: Action and reaction.

Flexible manufacturing systems have been designed with a high degree of automation to provide automatic tracking, alternative paths which are already set up and material handling to keep constant flow times or velocities. However, the outcome characteristic of predictable flow times, most important in terms of integration effects, can be obtained with other techniques besides computer control.

2.6 RELATIONS AMONG THE INTEGRATION VARIABLES

High levels of inventory, perfect balance or pure flexibility with universal, identical work stations can filter out integration effects in the production facility. In the past, integration effects have been eliminated with one of these techniques, but these are no longer economical solutions for most production facilities.

The Manufacturing Integration Model provides a means of measuring the trade-offs of inventory levels, of balancing capacity and of flexibility, which are needed to maintain an acceptable level of lost capacity due to integration (Fig. 2-24). If a production facility has

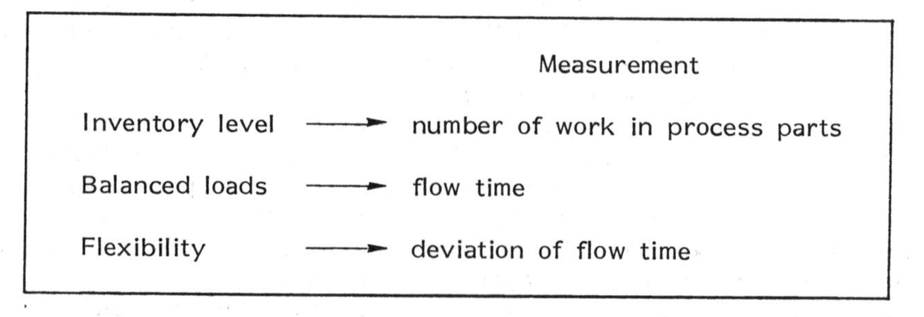

FIGURE 2-24 Trade-offs.

Primary goal:	(Productivity) Obtain desired production
Secondary goal:	(Minimum investment) Minimum inventory Unbalanced loads Dedicated processes

FIGURE 2-25 Primary and secondary goals.

★ Reduced set-up time

★ Increased flexibility

★ Balanced operations

FIGURE 2-26 MIM provides a means to study alternatives.

the objective to reduce inventory levels, MIM can instruct how much additional station availability or flexibility is needed to maintain a desired level of productivity. If a production facility involves a transfer line where less balance is desired, MIM can measure the increase in inventory or the amount of flexibility which will be needed to maintain a desired level of productivity.

In the application of MIM, each production facility will have some desired level of production. But to meet this level, many alternative secondary goals can be formulated (Fig. 2-25). These include a minimum inventory level, a desired flow time, a limit to the number of identical stations and available repair cycle for the stations. MIM provides a means of measuring and comparing the economics of these secondary goals (Fig. 2-26), and provides an up-front solution to a manufacturing goal, rather than an action–reaction approach.

2.7 FACTORY OF THE FUTURE

In review of these three variables, increases in inventory are not desirable solutions for the factory of the future. Such increases

Is an increase in capacity the most economical solution?

Net production = gross production — lost capacity

★ Lower inventory increases integration effects, lowers net production

★ Increase gross capacity to raise net production back to desired level

FIGURE 2-27 Increase in gross capacity effects.

have been used to solve integration problems in the past and have resulted in operatively expensive production facilities. Balancing is not applicable to most production environments and so its application for future factories will be limited. Increase in station availability by the acquisition of additional capacity has been the most common default means by which to maintain desired levels of production when inventories are reduced. However, this has the effect of converting variable costs (inventory) to fixed costs (capital/equipment) and means that the production facility will be less responsive to price variations. As a result, additional capacity to offset integration effects will only be cost-effective where price controls are effective (Fig. 2-27).

In terms of future manufacturing environments, flexibility is the one of the three variables which has the greatest potential. Certain levels of flexibility provide consistent velocity of part flow. This permits low inventories, minimum capacities and unbalanced operations. But its greatest benefit is in providing an economic approach to alternative manufacturing strategies so that the appropriate level of inventory with a realistic degree of flexibility can be obtained.

Flexibility also provides an umbrella for the implementation of many new technologies. Techniques for grouping like parts, for designing process which can be made flexible, design parts which can be produced flexibly and, of course, the new technologies of shop floor automation. In these forms, the only technologies which will be beneficial for the factory of the future will be those that offer increases in flexibility.

3

Flexible Manufacturing: The New Technology

The Manufacturing Integration Model (MIM) provides a means by which the trade-offs of a manufacturing strategy can be evaluated. These include the relationship between inventory level, balance loadings, station availability and flexibility. Chapter 2 briefly stated the trade-offs of each variable in MIM and its relation to integration effects.

This chapter deals with the relationship of flexibility to other MIM variables. Flexibility offers the ability to maintain lower inventory or reduce balancing. For this reason, flexible manufacturing should be thought of as an alternative technology for the production process.

3.1 INTEGRATION EFFECTS

The new technology is derived from MIM, which is based upon a view of the manufacturing environment. This view focuses on the integration effects due to the operation of the factory. There are two types of integration effects (Fig. 3-1). The first is defined as the amount of production lost when an available station is idle because there is no material for it. The in-process inventory has been depleted for the operations this station can perform and it must remain idle until all parts complete previous operations (Fig. 3-2). The second type of integration effect is defined as the amount of production which is lost due to a station being blocked by a part which cannot depart because there is no place for it to go. The in-process inventory here has filled the available space (Fig. 3-3), and now parts must remain at the station until a place becomes

★ Station is idle because of lack of material

★ Station blocked; unable to clear work area

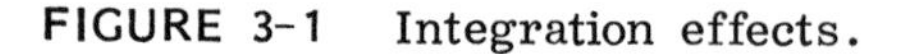

FIGURE 3-1 Integration effects.

available. Understanding both of these definitions for integration effects requires the following view of the manufacturing process.

Manufacturing can be viewed as a collection of parts, each having specific requirements or demands for stations. These demands materialize as the need to travel through a series of operations for which a number of different stations will be required. As a part travels through each required operation, the characteristics of each station will influence the part's flow.

The stations will continue to perform operations on parts as long as there are parts waiting (or arriving just-in-time), as long as the station has not yet failed and as long as it can free itself of completed parts. If these three conditions do not exist, the station is unable to continue operations and thus the production rate is lower.

FIGURE 3-2 Depleted in-process inventory for station operation.

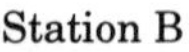

FIGURE 3-3 Filled in-process inventory for station operation.

- ★ Job shop
- ★ Transfer line
- ★ Flexible manufacturing

FIGURE 3-4 Major strategies of manufacturing.

Most capacity planning procedures include allowances for station failures by using efficiency factors. During a typical eight hour shift, there will be operator breaks for lunch and coffee and interruptions in the station due to a malfunction of its components, including tools. But these interruptions are easily observed and appropriate efficiency factors can be established which adjust capacity to realistic levels of expectation. In an eight hour shift, there may be a half an hour for lunch, some tool breakages and maybe one electrical problem on the station, and combined, they might account for a total of one hour when the station is not operating during the eight hour shift. This represents an efficiency factor of seven-eights of an hour or 87.5%. In planning the production capacity for this station, its total requirement cannot exceed 87.5% of the time available in the production period.

Efficiency factors will range from values as low as 70% to as high as 90%. These variations will be due to the nature of the process itself and labor environment. Of the three causes of integration effects, the capacity adjustment for interruptions in station availability is the one most often used. The lost capacity of a station being available with no parts and a station blocked by a part which has no place to go has not been quantified or even considered in many planning processes. The reason for this will become clear when each of the major strategies of manufacturing (Fig. 3-4), job shop, transfer line and flexible manufacturing, are presented from the viewpoint of integration effects.

3.2 JOB SHOP STRATEGY OF MANUFACTURING

The job shop was the first strategy of manufacturing and remains the most common today. Its characteristics consist of a variety of parts which require similar operations, but no single part with sufficient requirements justifying the purchase of special stations. Therefore, the job shop will contain general purpose stations, each with the capability to perform operations of one type. For example, a variety of parts will need some surface flatness. For this a milling station will be used, but this station will be used for a variety of parts.

The job shop can be characterized as having relatively low production volumes with a high variety of parts. To accommodate these characteristics, general purpose stations will usually be found in the job shop. Given these characteristics, the operation (production control) of the job shop creates its own environment.

3.2.1 Operation Environment

The operation environment of the job shop (Fig. 3-5) will have intermittent or batch production of each part. The frequency and size of these batches will be determined by lost capacity due to set ups, cost of the part and production requirements. Because a variety of parts will be produced using common stations in batch quantities, the work load will not be balanced for all stations. Some stations will require use at 90% efficiency, where others might only require 50% efficiency. This work load will also change as production requirements and more flexibility are incorporated into the types of operations a station can perform.

3.2.2 Integration Effects of the Job Shop

As in any manufacturing process, parts must flow through a series of stations to become a completed part. These stations will be going through their own cycles of operations, breakdowns, repairs and availability. As each of the stations fails, it will interrupt the flow of parts. This interruption may cause a station which can perform a subsequent operation to run out of work; that is, integration effects will appear when the subsequent stations become idle because of a lack of parts. This will reduce the capacity of the subsequent station, eventually lowering the number of parts which can be produced.

★ High inventory

★ Unbalanced station loads

FIGURE 3-5 Job shop operation environment.

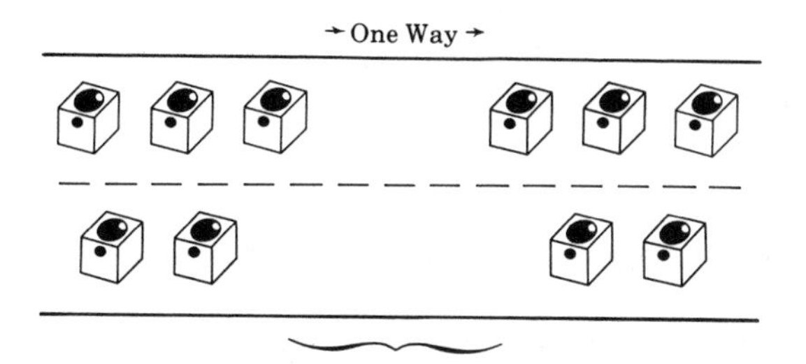

FIGURE 3-6 Window of inventory.

The flow of parts can be thought of as a continuous line, but when a station fails, this line is broken until the station resumes its operation. This break in the line is referred to as a window of inventory. If there is no accumulated inventory in front of the subsequent stations, this subsequent station capacity will be reduced by the size of the window of inventory. To eliminate the integration effect (Fig. 3-6), a window of inventory must be closed. It is closed by maintaining work in-process inventory between each station in the job shop. The size of these inventories will be consistent with the unbalanced loads of the stations and the length of inventory windows, which can be created from breakdown repair cycles of previous stations.

The second type of integration effect, where stations are blocked with no place to put parts, is not common in traditional job shops because parts are allowed to accumulate in any open space if needed. The only time this type of integration effect may appear in the job shop is when strict rules are established to prevent large buildups of in-process inventory.

Summarized, the integration effects in the job shop are managed (or controlled) by in-process inventory (Fig. 3-7). In fact, it is possible to raise inventory to such a level that integration effects

Integration effects are eliminated with inventory by never allowing the window of inventory to open

FIGURE 3-7 Job shop integration effects.

are completely eliminated. This will be particularly true for those manufacturing processes where the station acquisition and maintenance costs are high, relative to the cost of the parts.

3.2.3 Technology of Production Control in Job Shop

The lost capacity due to integration effects can be completely eliminated with in-process inventory. Production rates are then directly related to work in-process inventory levels. As inventory rises, the windows of inventory become smaller and capacity loss due to integration effects is lower. This increase in capacity translates into an increased production rate.

The problem of managing production in this environment is not related to capacity but rather to managing the part flow (Fig. 3-8). The high level of inventory results from very long flow times (elapsed time from first operation until completed part), which are highly variable. This variability creates the need to prioritize inventory so that part flow times are consistent with part production needs.

Tools which help solve this problem include queuing and scheduling theories. When the job shop is studied as part of an engineering curriculum, techniques of queuing theory (first in, first out, dynamic slack) and scheduling theory (Johnson's Rule, Jackson's Rule, Gantt Charts, PERT) are studied. For these parts which have dependent demands (their requirements depend upon the requirements of another part) Material Requirement Planning (MRP) has proven to be effective. One common characteristic stands out from all of these tools and that is that none of them deal with optimal use of the station. Instead, they all attempt to control the flow times of the parts by assuming a backlog of material for each station. These techniques will be effective as long as lost capacities due to integration effects do not exist. But as in-process inventories drop, integration effects will appear and these techniques will not be as effective.

Tools for inventory management include:

- ★ Scheduling theory
- ★ Materials requirement planning
- ★ Queueing theory

FIGURE 3-8 Inventory management.

3.3 TRANSFER LINE STRATEGY OF MANUFACTURING

The transfer line consists of characteristics which are opposite to those of the job shop. The transfer line was invented for the purpose of producing one class of parts. To accomplish this, stations with special characteristics are placed in a line. Because the path of the part is fixed, automated material handling systems are often found as links between these stations. The degree of automation in transfer lines is usually higher than job shops, but this should not be confused with its operation environment as necessarily being more highly automated.

The operation environment of a transfer line is derived from the continuous production of a part. The continuous production can be justified from the production requirements. This leads to the cost effectiveness of special stations customized for the specific operation of one family of parts. Because all parts flow through the same set of stations, the cycle times (operation durations) at each station must be about the same duration. If they are not, parts will accumulate for the longest cycle station and the capacity of the entire line will be determined by the longest cycle time.

3.3.1 Operation Environment

The operation environment of the transfer line (Fig. 3-9) will be the continuous production of a small variety of parts. The manufacturing process will be adjusted to provide uniform balance for all stations in the transfer line. This balance of cycle durations will provide the situation where the entire line will be stopped if even one station's service is interrupted. This operation is effective because all operations are balanced and the production rate is solely determined by the longest cycle operation. But when all operations are the same, it does not make any sense to accumulate inventory in this line. Therefore, when one station fails, the entire line stops. This procedure maintains the uniform balance of the transfer line.

★ Low inventory

★ Balanced station loads

FIGURE 3-9 Operation environment of the transfer line.

3.3.2 Integration Effects in the Transfer Line

Both types of integration effects are highly visible in the transfer line. The first type, where stations are idle because of a lack of material, is observed whenever one station fails along the line. In this case, all station operation is synchronized, so there is no chance for windows of inventory to appear. The nonsynchronous transfer line can have empty stations due to differences between cycle times.

The other integration effect where stations are blocked because the part has no place to go occurs in most transfer lines. Because of its high visibility and direct relation to lost station capacity, efforts are made to eliminate its occurrence and independent station operation.

The method for eliminating the integration effects within the transfer line is to balance the operations within the line (Fig. 3-10). This is usually accomplished by adjusting the process until the series of operations, which are required by a part, are all of the same duration. As long as the operations are balanced, integration effects can be eliminated.

3.3.3 Techniques of Transfer Line Control

When a transfer line has balanced operations, the integration effects are eliminated. Its production rate (or capacity) is then determined from the cycle time or the flow time of the part. But the flow time is inversely related to production. The shorter the flow time, the higher the production rate and vice versa.

The techniques of managing transfer line production address flow time just as in the case of techniques for job shops. But the techniques for transfer lines are directed toward the process itself instead of at inventory. Therefore, when studying transfer lines in an engineering curriculum, two techniques are presented: line balancing and the line of balance technique (Fig. 3-11). Both of these techniques are applied in the process engineering of the transfer line but are not effective in addressing operational problems.

The transfer line provides constant flow times and balancing eliminates integration effects. Because integration effects are eliminated with the balancing, there is need for in-process inventory

Integration effects are eliminated by maintaining perfect balance within station loads

FIGURE 3-10 Eliminating integration effects in the transfer line.

Tools for balancing include:

★ Line balancing

★ Line of balance

FIGURE 3-11 Balancing the transfer line.

between stations. For this reason, the entire transfer line can be treated as a single operation to manage material flow throughout the entire factory.

3.4 FLEXIBLE MANUFACTURING STRATEGY

The job shop can be defined as the manufacturing environment which consists of unbalanced operations and where high levels of inventory exist to reduce integration effects upon capacity. The transfer line can be defined as the manufacturing environment, consisting of balanced operations that reduce integration effects on capacity and low levels of inventory. Consequently, flexible manufacturing is defined (Fig. 3-12) as the manufacturing environment which consists of low inventory levels and unbalanced station loads or operations. (In these definitions, the degree of automation is not important; only the relative balance of work load across stations and levels of work in-process inventory.)

By definition, flexible manufacturing provides an environment where integration effects cannot be eliminated. If inventory is raised, the environment becomes that of the job shop. If operations are balanced, its environment becomes that of the transfer line.

From this definition, integration effects must be present in flexible manufacturing. The operation of a flexible manufacturing system

★ Low inventory

★ Unbalanced station loads

FIGURE 3-12 FMS operating environment.

Integration effects can be completely eliminated only when all work stations have identical capabilities

FIGURE 3-13 Quantify lost capacity due to integration effects.

will have stations which are idle because of material shortage and stations which are blocked because there is no place to deposit parts. Because these two situations can never be eliminated, the techniques for planning and controlling flexible manufacturing must be able to quantify the lost capacity due to integration effects (Fig. 3-13). The loss in capacity never needs quantifying in job shops or transfer lines because their effects can be eliminated through environment changes. The need to quantify integration effects establishes the need for a new technology. To start this new technology, a model must be constructed to quantify integration effects. The Manufacturing Integration Model (MIM) presented in Chapter 2 provides a framework for the development of new techniques and quantification of them. One method for quantifying integration effects has been established and is described in the next section.

3.5 TRIANGLE OF INTEGRATION

From the job shop, a relationship has been established between production rates and inventory levels (Fig. 3-14). As the inventory or work in-process level raises, production will increase because the

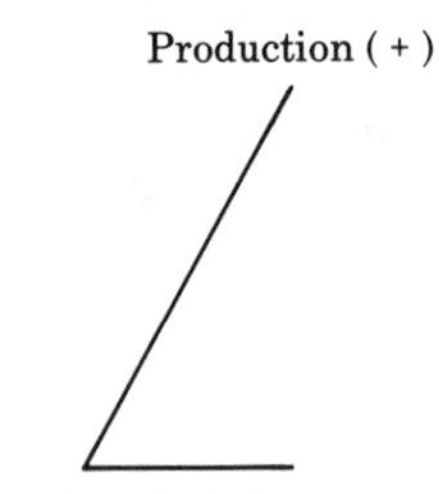

Increases in inventory yield increases in production

FIGURE 3-14 Production vs. inventory.

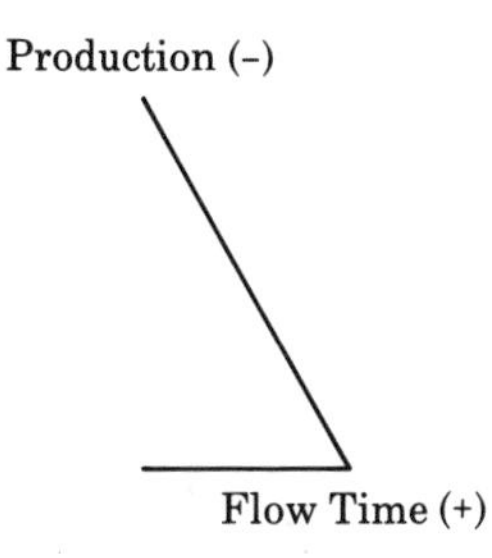

Increases in flow time yield decreases in production

FIGURE 3-15 Production vs. flow time.

windows of inventory become smaller. At this point in the model definition, the exact amounts of inventory which are needed to raise production any given amount are not important. What appears outstanding is that production and inventory are directly related.

The relationship between production rates and flow time (Fig. 3-15) has been established from the transfer line. As flow time decreases or parts move through the line at higher speeds, production increases. Production decreases when parts spend more time in the transfer line. So, production and flow time are inversely related.

As long as integration effects are eliminated or are assumed so, production is controlled either by controlling inventory in the job shop or balancing operations in the transfer line. However, in flexible manufacturing, integration effects are not eliminated and the triangle of integration (Fig. 3-16) provides a means to quantify these integration effects.

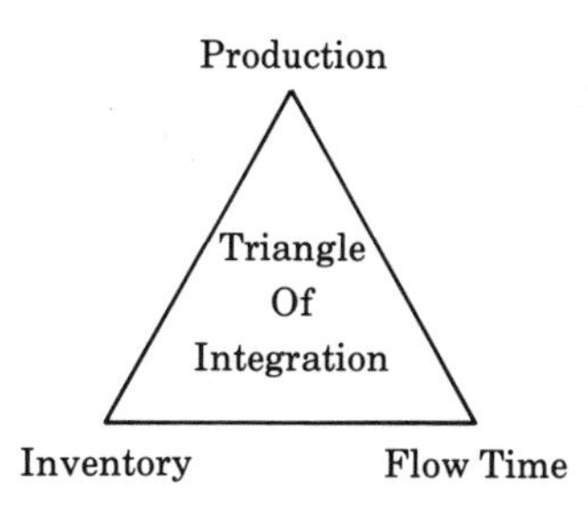

FIGURE 3-16 Triangle of integration.

Units of measure	
Inventory:	Number of parts in process
Balancing:	Flow time
Flexibility:	Change in flow time

FIGURE 3-17 Quantifying integration effects.

There are three variables which determine the amount of integration effects that result in a production process: inventory level, balanced loadings and flexibility. Qualifying the integration effects requires the ability to measure each variable (Fig. 3-17).

Inventory level is quantified by counting the number of parts which are active in the production process. This number can represent the number of individual pieces, a batch of parts or a container.

Balanced loadings can be quantified by the use of flow time. The use of flow time to measure the balance within a production facility is derived from the transfer line. The capacity of these stations is captured within the specific amount of work and when this work is equally balanced across the 10 stations, integration effects will be eliminated. Also, this will be the shortest flow time the part can have in the transfer line. From this characteristic of the transfer line, flow time can be used to measure the degree of balance within the line. This concept has been generalized such that flow time provided a means to measure the balance between station loads in any type of production facility.

The Manufacturing Integration Model suggests that *flexibility* can be measured by the variability of flow time. A process with greater degrees of flexibility will provide less variability to the flow time. As the station becomes more dedicated to operations and set up time (conversion from one operation to another) increases, flow time will have greater variability.

In these terms, the average flow time can measure the balance loadings between operations and the variability of flow time can be used to measure the degree of flexibility in the production facility.

The qualification of MIM can be achieved with the measurement of three variables: work in-process inventory levels, flow time and station availability. An increase in production might require an increase in inventory because the two are positively related to one another. However, as we increase inventory, higher levels of congestion will appear and increases in flow time will cause production to decrease.

The *triangle of integration* indicates that changes in production are related to both inventory changes as well as changes in flow time. For example, an increase in inventory (number of parts) in the manufacturing system will provide some increase in production. But the amount of this increase will be offset (adjusted lower) by the decrease in production due to an increase in flow time. Therefore, the marginal change to production is always a function of inventory and flow time.

The triangle of integration establishes the model of parameters which establish the identification of integration effects. The next step is to provide a technique which will quantify this relationship for various levels of inventory, flow time and station availability.

3.5.1 The WIPAC Curve

The Work In-Process Against Capacity (WIPAC) Curve (Fig. 3-18) is a two-axis graph showing the relationship between inventory levels and production rates.

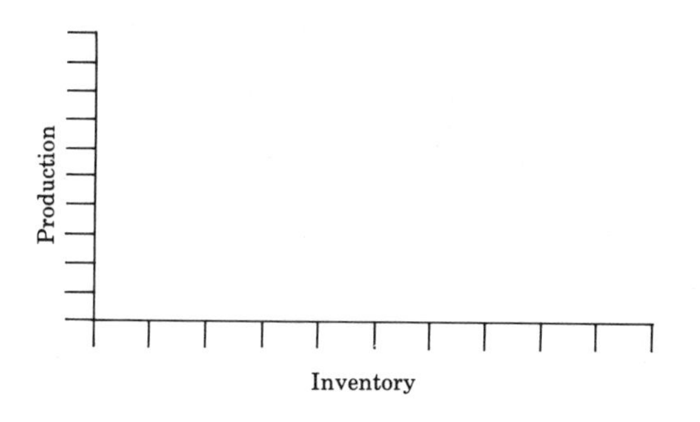

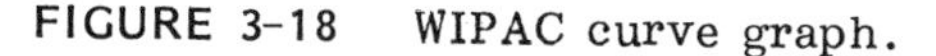

FIGURE 3-18 WIPAC curve graph.

The horizontal axis represent the work in-process or inventory which is quantified in terms of the average number of parts which are *active* in the production process; that is, the part is able to move freely throughout the production process as stations are made available. The vertical axis represents the production that can be quantified in either units of parts per hour or in terms of station utilization. Station use and parts per hour are identical measures but expressed in different units of measure. Whichever is not easily obtained can be used to establish the initial scale and determine the other from it.

3.5.2 Characteristics of the WIPAC Curve

The shape of the WIPAC Curve can be used to establish specific relationships between the effects of inventory, the effects of flow time and station availability upon production.

From MIM, increases in inventory reduce integration effects and the production rates must increase as well. Therefore, when graphing inventory against production, we expect the curve to be upward sloping. However, as inventory levels increase, flow time must increase as well. How much it increases (its degree of variability) is determined from the balance and flexibility of this production process. But as long as it increases, this will have a negative or downward push on production rates. Thus, it is this effect that causes the WIPAC Curve to bend.

The relation of inventory and flow time to production rates can be discussed in four cases. These are:

Case 1: Curve has constant slope, curving up and to the right. Whenever this characteristic (Fig. 3-19) is observed along the WIPAC Curve, an increase in production is directly proportional to the increase in inventory. Flow time must be constant along this line or its increase has insignificant effect upon production.

Case 2: Curve has decreased slope, curving up and to the right. Whenever this characteristic (Fig. 3-20) is observed along with the WIPAC Curve, production increases because the positive effects due to increases in inventory are greater than the negative effects due to a flow time increase. Therefore, flow times are increasing as inventory increases which causes the curve to bend downward.

Case 3: Curve is flat with a slope of zero.
Whenever this characteristic (Fig. 3-21) is observed along with the WIPAC Curve, increases in production due to increases in inventory are equal to a decrease in production due to increases in flow time. Therefore, production does not change.

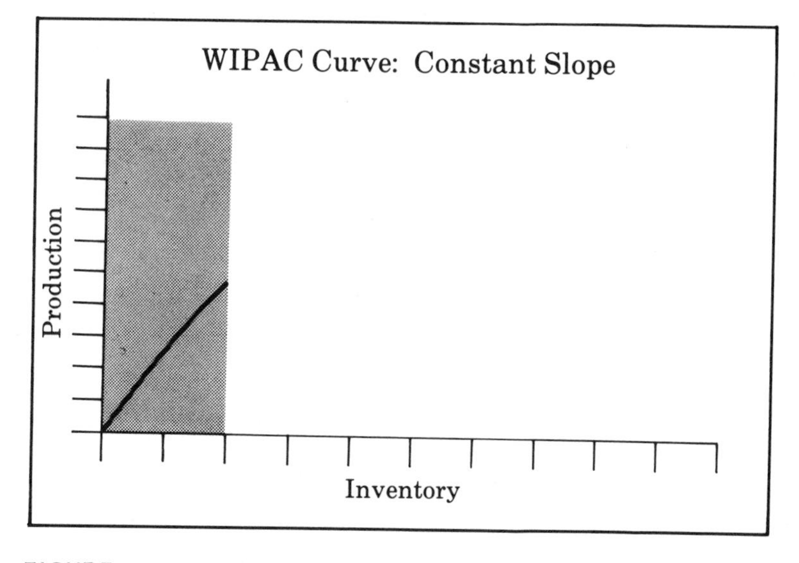

FIGURE 3-19 Constant slope.

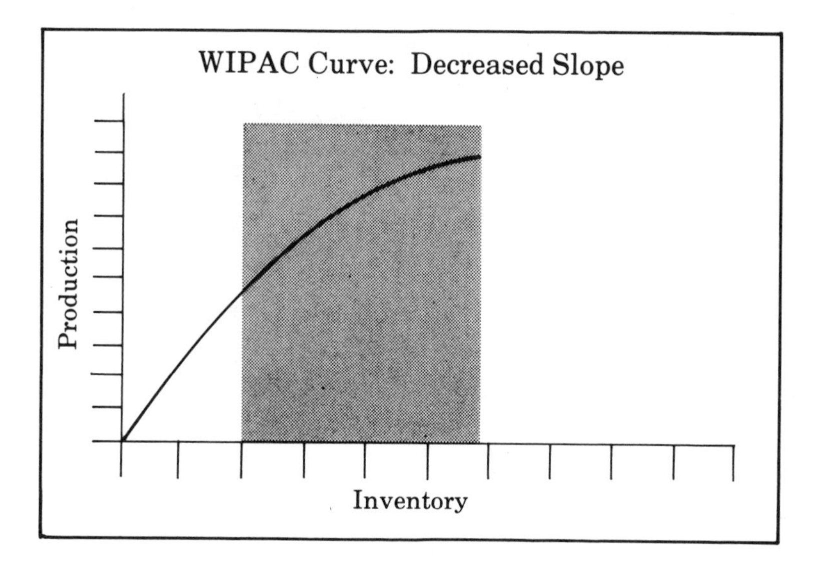

FIGURE 3-20 Decreased slope.

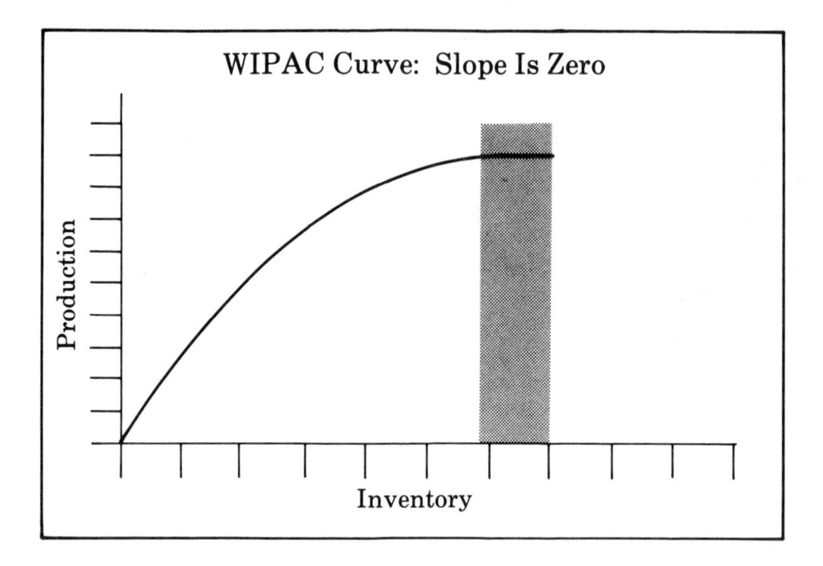

FIGURE 3-21 Zero slope.

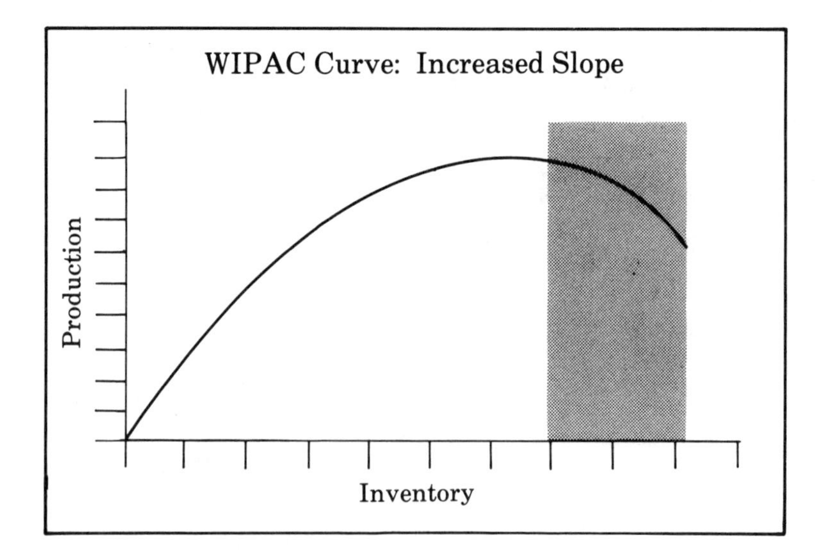

FIGURE 3-22 Increased slope.

Case 4: Curve has increased slope, curving down and to the right.
Whenever this characteristic (Fig. 3-22) is observed along the WIPAC Curve, production is decreasing when inventory is increasing. Therefore, the decrease in production due to the increase in flow time is greater than the increase in production due to the increase in inventory. This establishes the relation of inventory and flow time to production, but the relationship of station availability must be explained as well. Station availability is represented as the maximum production capacity which can be attained. This becomes the ceiling for the WIPAC Curve. As stations fail and repair, this ceiling constantly moves and the WIPAC Curve responds as if a weight were being raised and lowered on a spring.

Another characteristic of the WIPAC Curve is its maximum height. When there are no integration effects, the WIPAC Curve will touch the ceiling established by the station availability (Fig. 3-23). In these situations, production is determined directly from station availability. This characteristic can be achieved through perfect balancing or high inventory levels as described for a transfer line on job shop manufacturing strategy.

However, when integration effects do exist, the height of the WIPAC Curve will be less than the ceiling established by the station

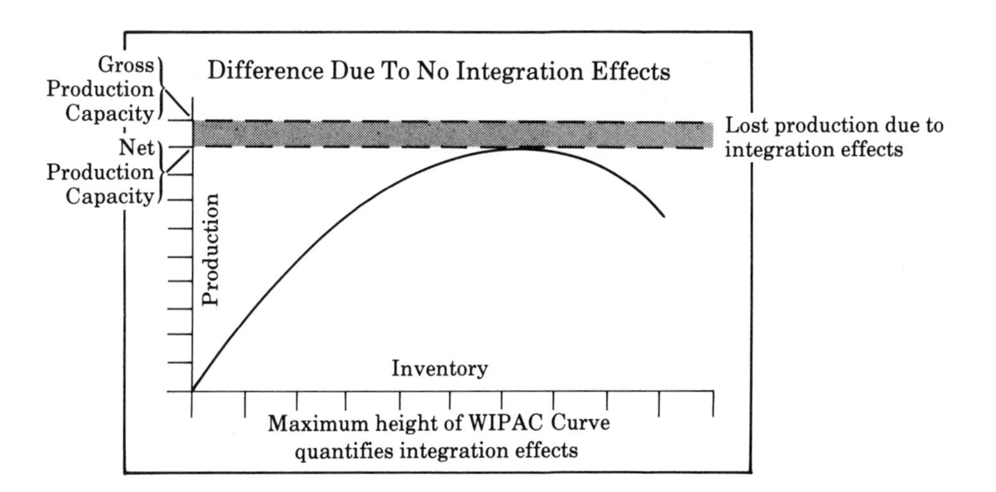

FIGURE 3-23 No integration effects.

availability. The difference between these points is the lost production rate due to integration effects. Thus, the implementation of the WIPAC Curve provides a means to quantify the integration effects of inventory level, balance loadings, station availability and flexibility.

3.5.3 Implementation of the WIPAC Curve

The WIPAC Curve provides a means to graphically quantify the relationship of inventory level, flow time, station availability and production ratio. Production can be mathematically derived from station availability by use of capacity planning techniques. Production rate can be derived from flow time by use of operation durations. However, inventory and flow time are not independent variables, but rather are related to one another.

Some methods have been developed which establish mathematical relationships between an inventory level and flow time. With these models, the WIPAC curve can be constructed from pure mathematics (Fig. 3-24). The first such tool was developed by Dr. Jim Solberg of Purdue University, entitled CAN-Q. CAN-Q was first available in 1977 and used a queuing theory and stochastic processor to estimate queue times for parts in flexible manufacturing systems.

Once a queue or wait time can be estimated for each operation, the flow time for each part can be derived. This then establishes a relationship between a number of parts, expected flow time and a production level.

A second technique for estimating the WIPAC Curve comes from a doctoral dissertation of Dr. Viedinger in Berlin, West Germany. His technique also uses a stochastic process to estimate lost production capacities due to increases in flow time.

In both of these techniques, the mathematics are complex. However, CAN-Q can run on an IBM pc and produce results within 30 minutes of computer time. The techniques' strengths are quick

1. CAN-Q [approximates]
2. Fraunhofer technique [approximates]
3. Computer simulation [exact]

FIGURE 3-24 Techniques for generating the WIPAC Curve.

results but their weakness is in the assumptions which are necessary to establish the queuing theory probabilities. As long as assumptions are not violated, they will produce accurate results.

Because both of these methods require some assumption or heuristics, they can only approximate the WIPAC Curve. There are no mathematical means at this point to precisely relate inventory flow time and production. Therefore, the best technique for precisely naming the relationship for the WIPAC Curve is computer simulation. In fact, each point of the curve has been generated by a simulation run.

Part III describes a step by step procedure for flexible manufacturing design and evaluation, and the procedure for generating the WIPAC Curve.

3.5.4 Application of the WIPAC Curve

From the description of implementation, the WIPAC Curve generation requires the use of both mathematical models and computer simulation. As this is explored further in Part III, you will find that the shape of the curve is also dependent upon the part mix. As the part mix changes, so must the shape of the WIPAC Curve. Also, the station availability is a constantly changing variable. This constant motion has a similar effect of a weight being raised and lowered on a spring. However, the spring does not come in contact with the weight because of integration effects; yet its proximity varies. From this brief description, much expertise is needed in use of simulation, evaluation of results and in the study of a dynamic problem.

In the case of the WIPAC Curve, it is important to identify the actual points of the curve and to study how it changes relative to the dynamics of the production process. It is evident that there is a need for an expert system to assist not only in the preparation and generation of the WIPAC Curve but also in the evaluation of its results. MIM and the WIPAC Curve provide a framework upon which an expert system can be based, and has become a topic for current developments.

3.6 CONCLUSION

Flexible manufacturing is an outgrowth of existing manufacturing technologies, but its design and control has not been an outgrowth. Flexible manufacturing provides a low inventory environment with unbalanced operations unique to the conventional production environment. There are substantial benefits with this environment, but the technology of job shop or transfer line falls short of providing tools to realize these benefits.

The triangle of integration provides a model from which greater understanding of flexible manufacturing can be gained. The ability to study flexible manufacturing systems from an integration effects point of view is essential because integration effects cannot be eliminated and will influence the production capacity. The WIPAC Curve provides a simple, graphical means to quantify the triangle of integration model. This curve can explain why many flexible manufacturing systems today are not achieving maximum benefits and provides the information from which flexible manufacturing can be made into a reliable, new technology of manufacturing strategies.

4

Flexible Manufacturing System Hardware Components

The hardware components in the FMS include equipment from every area of manufacturing, including machines, robots, inspection stations, storage facilities, material handling, deburring, chip removal and manual work areas. The first question which might be raised is "Why are manual work areas on this list of advanced manufacturing techniques?" Because the FMS of today has not yet reached 100% automation, operators still remain in order to carry out complex operations where automation is infeasible, too costly, or unreliable. Significantly, the FMS is a manufacturing system which intersperses manual operations amongst the most advanced factory equipment. This is a normal component of the FMS definition. The automatic operation can fall into two general categories: machining or assembly-type operations.

Below is a detailed description of each hardware component area listed above. Individual characteristics are not intended to present the merits of each component, but rather to explain how components fit into the FMS definition.

4.1 PALLETS AND FIXTURES

The fundamental component which allows for integration of machines, material handling and in-process storage facilities is the use of palletized parts. The integration is accomplished by uniform material handling and deterministic part organization. The palletized part (Fig. 4-1) consists of three components: (a) the part, (b) the fixture and (c) the pallet. The pallet is a steel disc or square with slots in its surface. These slots are used to fasten the fixture to

FIGURE 4-1 Part, fixture and pallet. The round disk with slots is the pallet. The part is the large object containing the holes and the fixture is the clamps which position the part into a predefined orientation. (Courtesy of Giddings and Lewis Corporation.)

the pallet. The fixture is configured for a single part or, at most, a family of various parts, and provides a fixed orientation of the part at all times.

It is possible to change fixtures or pallets, but the fixtures must be precisely located on the pallet to ensure consistent part orientation. Because of this need for precise location of fixtures or pallets, often the fixture is permanently attached to the pallet. This prevents wear and lost accuracy caused by frequent fixture changes. It also eliminates the need for a trial run of a single part to check for accurate part orientation. Therefore, if additional fixtures or new fixtures are required, pallets are often purchased as well.

The two factors which contribute to the high cost of pallets and fixtures are physical hardware and precise accuracy. Often, accuracy is the largest contributor to cost and does not only apply to the individual pallet/fixtures. Consistent accuracy is required for all pallet/fixtures which are designed for the same part. Every fixture/pallet must position the parts in the same axis so that all surfaces of the part are parallel with all corresponding surfaces of identical parts. Besides the requirements of parallel surfaces, all parts must be positioned at the same height above the pallet. Each fixture of the same type might have to be identical to ensure part quality. However, offsets can often be used to adjust for minor variations in each fixture/pallet.

Such offsets are minor adjustments to the axis of the machine so that the distance from the part surface to the tool can be made identical for each fixture/pallet. These offsets can only adjust for errors in a single axis; in some cases, an offset can be used for each of the three axes. When three offsets are used, it is possible that, taken in combination, they will result in a lack of squareness. Therefore, it is common to allow for a single offset, but the fixture must be precise in the other axes and in parallel part orientation. Chapter 5 contains further discussion in the use of offsets in machine adjustments.

Once the orientation of the part is determined, the sequence of machining operations is also known. These fixtured parts will then be delivered to a variety of stations in order to complete the necessary operations. The most common type of station in the FMS is the machining center.

4.2 MACHINING CENTERS

The machining center consists of a column, work table, tool storage and tool changer. Machining centers can be either vertical, horizontal or lathe, depending on the orientation of the spindle. Figure 4-2 is an example of a horizontal machining center with a visible column, which contains the spindle and the work table.

4.2.1 Three-Axes Machining

The column contains the spindle, which can be moved in three axes of motion, x, y, z (Fig. 4-3). The "z" axis of motion is arranged parallel to the axis of the machine spindle. "Z" represents the movement of the spindle into and out of the work piece, accomplished by moving either the spindle or the work table. The "x" axis of motion is horizontal and parallel to the work table. "X" represents the horizontal motion of the work table relative to the spindle. This can be accomplished by moving either the part or the work table. The "y" axis of motion was selected to complete the cartesion coordinate system. "Y" represents the vertical motion of the spindle relative to the part and is always accomplished by moving the spindle. Figure 4-4 shows a vertical machining center where the "x" motion is the same as in the horizontal machining center but the

FIGURE 4-2 Horizontal machining center. The spindle is oriented in the horizontal plane. (Courtesy of Kearney and Trecker Corporation.)

● x-axis travel—120" and 156". Extended runway travels to 31'
● y-axis travel—36" and 42"
● z-axis travel—24"
□ Table size—24" or 32" wide
132" or 168" long

FIGURE 4-3 The three axes of machining. "X" axis is left to right, "y" axis is along the spindle and "z" axis is up and down. (Courtesy of Giddings and Lewis Corporation.)

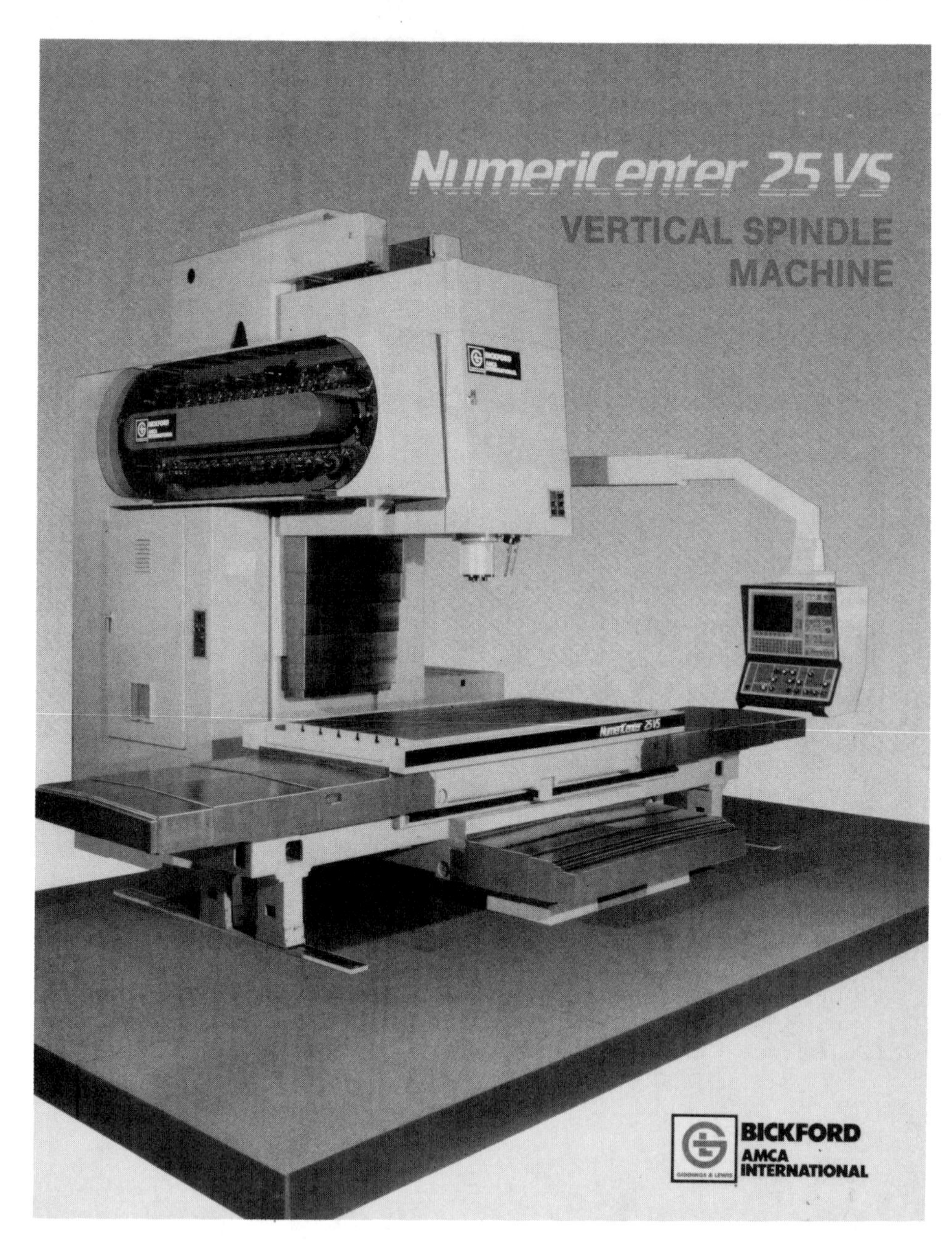

FIGURE 4-4 Vertical machining center. The spindle is oriented in the vertical plane. (Courtesy of Giddings and Lewis Corporation.)

"y" and "z" axes are reversed. With the vertical spindle, the "z" motion becomes "y" in the horizontal. The length of travel in x, y, z determines the size of the part that can be machined. Three-axes motion (i.e., x, y, z) results from characteristics of the columns by spindle movements. Additional four- or five-axes motion is gained from movements within the work table. The work table is the resting area for the part while it is being machined. A flat machining surface, called a fixed table, is used for three-axes machining.

4.2.2 Four- and Five-Axes Machining

The four-axes machining center which consists of x, y, z axes also includes a "b" axis, which is a rotation in the table. Figure 4-5 shows a vertical turning center with the rotary work table. This rotation of the work piece allows for contouring machine operations, such as smoothing curved surfaces, accomplished by positioning the spindle with a tool in some x, y, z and then rotating the part. This allows the machine access to all sides of the part.

The five-axes machining center includes the x, y, z, b axes along with the "a" axis, which is a tilting of the table towards the column. With the spindle positioned in some x, y, z and by tilting the table in the "a" axis, compound angle holes and surfaces can be machined.

The machining center has flexibility in performing many different types of machining operations, from simple hole drilling to five-axes contouring. Some of the many operations which can be done include drilling, tapping (putting a thread into a drilled hole), boring (increasing the size of a hole), milling (smoothing the surface of a work piece) or turning (a process where the work piece moves and the spindle is fixed in position). However, to take advantage of this diverse set of machining operations, the machining center must be able to change tools.

4.2.3 Tool Storage/Tool Changers

Tools are held in tool holders (Fig. 4-6), which provide a uniform means for handling and locking the tools into the spindle. These tools are housed in the tool storage, which is a circular disc or chain containing tool pots (Fig. 4-7). These pots are designed to hold one tool holder, containing a tool.

These tools can be automatically taken out of the storage and inserted into the spindle by a robotic arm. This is accomplished by the tool changer. The removal of the tool from the spindle and the

FIGURE 4-5 Vertical turning center. The part and table turn while the spindle remains in a fixed position. (Courtesy of Giddings and Lewis Corporation.

insertion of the next tool is accomplished in a single motion (Fig. 4-8) to minimize the time delay for the tool change. After the tool change, the tool which is being replaced is put back into the tool storage. Two schemes are available for replacing a tool into the tool storage. The simplest method is to return the tool back to its original pot number, thus allowing for its identification by tagging

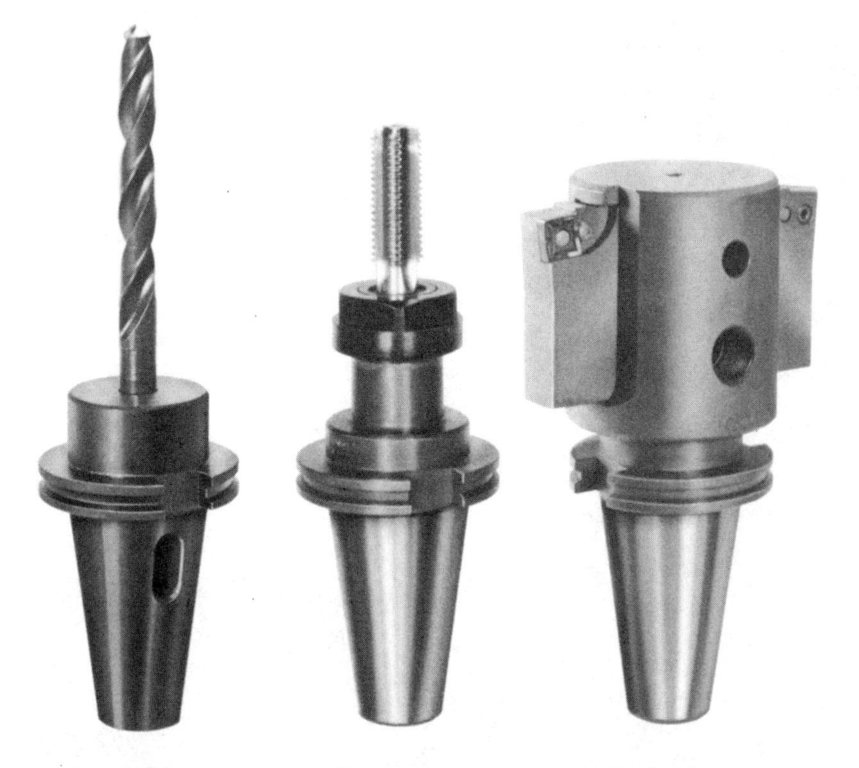

FIGURE 4-6 Tools with tool holders. The tool holders permit uniform handling and spindle clamping of any shaped tool. (Courtesy of Giddings and Lewis Corporation.)

it with a pot number. However, the program directing motions of the machine expects the tools to be in specific locations, and any change will result in the machine stopping. Therefore, this method might cause delay in the time needed to reposition the tool storage to the desired pot. The second and more efficient strategy is to place the used tool into the pot previously used by the replacement tool. The advantage of this is that no repositioning of the tool storage is required before the tool can be put away. In this case, however, the control must be dynamically able to keep track of this

FIGURE 4-7 Horizontal machining center with tool storage chain. (Courtesy of Kearney and Trecker Corporation.)

FIGURE 4-8 Tool charger in action. Tool has been removed from spindle and is being replaced with another. (Courtesy of Giddings and Lewis Corporation.)

Contouring Head

FIGURE 4-9 Contouring head type of tool with tool holder. (Courtesy of Giddings and Lewis Corporation.)

tool as well as the pot in tool storage which is housing it. This is complicated because some tools have large diameters which prohibit them from occupying just any position in the tool storage. For example, the contouring head in Fig. 4-9 might have a diameter of six inches but its shaft narrows to fit into a single tool holder. When placed in the tool store, only a single pot is needed. But because the tool hangs over the holder, it is impossible to use the neighboring pots for other tools. Figure 4-10 shows a tool magazine containing overhanging tools. For these reasons, it is not always possible to replace one tool with another in the same tool pot.

Tool identification allows tools to be placed and replaced anywhere in the storage. Each tool holder is specially coded according to its type, and as the tools rotate in the storage, they pass a sensor which can read and identify each code on the tool holder. Therefore, when a specific tool is called for, the tool storage rotates until it is located. This feature also permits tools to be manually added and replaced without the restriction of having a specific tool in a specific pot.

This discussion of tool identification is an illustration of the trade-off between flexibility and complexity of operation. The simplest mode of operation is assigning tools to designated pots, but this limits the ability to add new tools. To increase the

FIGURE 4-10 Overhanging tools. Some tools are spaced apart because tool diameters are larger than the spacing in the tool storage. (Courtesy of Ex-Cell-O Corporation.)

flexibility, tool identification is required. It might appear to be operationally simpler to allow any tool to be placed in any pot, but because of the dynamic nature of operations, recovering from contingent problems is more complex.

4.3 ROBOTS

Due to the diverse nature of robots and their flexibility in motion, they have various forms of application in flexible manufacturing.

4.3.1 Pick and Place Operations

The most common application of robots within FMS are pick and place type of operations, where point to point controlled devices are sufficient. These applications include tool changing, load/

FIGURE 4-11 Robots as tool changers: an overhead gantry robot is used to replace multiple spindle heads. Individual tools are replaced with the conventional tool changer at the left. (Courtesy of Ex-Cell-O Corporation.)

unloading unfixtured parts into work tables. As shown in Fig. 4-11, a robot can be used to replace tools in the spindle of a machining center. This application is much the same as the tool changer mechanization except that the tool store will not reposition. The robot will pick and place tools directly into designated areas in the store, and in this application it is possible for one robot to change tools to be shared by several machines. But when two machines require use of a single tool, one will be delayed until the tool becomes available. In the application of part handling either for fixture loading or work table loading, the robot must be able to hold the part at a point which permits placing it into a fixture and activating the clamping sequence. In some cases, to facilitate robot handling of the parts, specially designed adjustments to the part might be necessary. These handling enhancements might be additional threaded holes, tabs, or pockets which would not be needed for manual part handling and fixturing. The cost of preparing and machining these handling features to the part must be compared to the cost of conventional part handling.

4.3.2 Contouring Operations

A second major application area is for robots in contouring type operations. These include welding, limited machining, deburring, assembly/disassembly and inspection. In the case of welding, robots have proven to be reliable, effective and efficient. However, in the other areas such as machining, deburring and inspection, robots' limited accuracy and repeatability limited their application. In addition, whenever a tool change is required, such as in deburring, the cost of a robotic change is almost as expensive as that of a three-axes machining center. With the reduced accuracy of the robot as compared to the machining center, it is possible that simple operations which require multiple tools are most efficiently performed in machining centers.

4.3.3 Assembly/Disassembly

The use of robots within FMS is much more likely in assembly and disassembly. The robot will be the central component in any flexible assembly system. Again, in islands of automation, the robot has been proven effective for the assembly of small parts and printed circuit boards.

One final area of application for robots is in the inspection of parts.

4.4 INSPECTION EQUIPMENT

Since the FMS is a closed system, it is necessary to provide some means to monitor the quality of the operations being performed. This monitoring can take place in many different places and by many different components.

4.4.1 Coordinate Measuring Machine

The most obvious type of inspection equipment is a coordinate measuring machine (CMM) (Fig. 4-12). This machine can be programmed to probe a piece part and identify depth of holes, flatness

FIGURE 4-12 Coordinate measuring station for palletized parts. Parts pass through the station and are positioned for the inspection process. (Courtesy of DEA.)

FIGURE 4-13 Probing machine center. This probe has been inserted into the spindle and will send signals to the control as to any deflection it encounters. (Courtesy of Giddings and Lewis Corporation.)

of surface, concentricity, and perpendicularity. Special requirements usually include constant temperature congruity between environment and piece part. Also, because of the slow movement necessary to precisely measure surfaces, the inspection time is usually long compared to the actual machining.

4.4.2 Robots/Probing Machining Centers

Alternative types of inspection equipment other than the special CMM station include the use of robots or probing machining centers. These machines inspect equipment in the work center by inserting a probe into the gripper or spindle and then moving the probe to contact the work piece or fixture. Figure 4-13 shows a probing machining center. Unlike other tools, the probe has a unique characteristic which enables it to send a signal to the control when it comes in contact with a surface. This is not as accurate as the CMM and takes a relatively long time to complete.

The advantage of inspecting in the work center is that adjustments can be made prior to making an out-of-tolerance cut, whereas the CMM always inspects after the machining process. The disadvantage of using probing is that it requires more machine time and the degree of accuracy is not as high as with CMM.

4.5 CHIP REMOVAL SYSTEMS

Chips are defined as the pieces of metal which have been removed from the piece part. There are two methods of removing these from the work areas: chip conveyor to collection box or an in-floor flume system with a centralized collection area.

4.5.1 Chip Conveyor/Collection Box

The first method, chip conveyor to collection box, allows the chip to fall into a pit under the work table. A conveyor is then used to drag the chips up an elevator where they are deposited into a collection box (Fig. 4-14). In this type of collection, a separate collection box is needed for each station.

4.5.2 Centralized Flume System

The second type of chip removal and collection system is a centralized flume system, discussed further in Chapter 12. In this system, chips drop into a trough or flume which runs under the table. A fluid is pumped through the flume and the chips are deposited into a centralized collection area. Chips from all stations are then deposited into a single area.

FIGURE 4-14 Chip conveyor collection. Chips are conveyed from the bottom of a machine to the collection box by a drag chain conveyor.

The advantages of the central flume system are its aesthetics for the installation. The operator area is cleaner, since only a few chips remain in the work area, and there is no need to monitor individual station collection bins. The disadvantages are its expense in installation and operation. Operation expenses include power to pump the fluid and cost of fluid replenishment through evaporation. This evaporation can also create excess humidity and cause problems with electrical components. Another disadvantage is the inability to separate chips. For example, it is possible to have some steel parts and aluminum parts in the same FMS. But a centralized system will mix these chips together, and unless they are separated by another means, their salvage value will decrease. When chip separation is important, the chip conveyor for each machine has the advantage of being able to avoid contamination of mixing chips.

4.6 MANUALLY OPERATED STATIONS

As stated above, the operator is an important component in most FMS and plays an important role in bridging the gaps between various machinery. This section is not intended to describe the procedures or the responsibilities of the operator, but rather to present the type of operator controlled work stations as they might exist in an FMS. Chapter 13 contains a detailed discussion on FMS operator roles.

Three general types of work stations have been used in FMS. These are pallet load/unload areas, tool preparation and checks in machining accuracy or stress relief.

4.6.1 Pallet Load/Unload Areas

The most common FMS work station is the pallet load/unload area. This station is usually a pallet stand (Fig. 4-15) or an open area

FIGURE 4-15 Manually operated load/unload stand with pallet and fixture part. (Courtesy of Detroit Diesel Allison—Steve Fox.)

where all operations take place, with the transport system serving as the work table. In either type of work table, the objective is to take piece parts and position them onto the pallet. If the parts are under one hundred pounds and the clamping of the part is simple, robots can be used to load and unload pallets. However, in most cases, these parts require complex fixturing and clamping sequences, which currently prohibit robotic application.

These work stations can be separate stands for pallet load and unload, common areas for load and unload, or areas where the pallet remains in the material handling system. These areas are not restricted to load and unload but can also be used for reorienting the part for machine access to hidden surfaces. Again, reorientation of the part can be done at designated areas or in common areas for all activities. Another type of manual load/unload area employs the use of specific part types dedicated to specific areas. Often a part will require multiple passes through the system, each pass requiring a different orientation and a different type of fixture. To simplify the load/unload procedures, first sequence parts are separated from second and third sequence parts. Such simple operation restrictions reduce the otherwise dynamic nature of loading/unloading multiple sequence parts.

4.6.2 Tool Preparation

Manual areas are also designated for tool set up and delivery into and out of the FMS. Often the operators are responsible for setting the necessary tools. The tool setting area can be interfaced directly to the material handling system for automatic tool pick up and delivery. These tools are kept in either a tool cube or replacement disc. When the tools are delivered to the station, they must be replaced one at a time or the entire disc is swapped. In most applications where tools are automatically replaced at a machine, a second tool changer is used to avoid use of the work table and spindle. This allows the station to continue its current activities without interruption for tool replacement. The control and coordination of machine activities and tools is complex and is described in the next chapter.

4.6.3 Part Stress Relief

A third type of manual station is part stress relief. After a machining operation or even in the middle of one, the part might need to be unclamped and then reclamped to relieve twists and other stresses which result from machining a fixtured part. These stations could simply be hold positions in the material handling system where a

FIGURE 4-16 Part clamping in hold position. Carts with pallets arrive and wait for operator response. (Courtesy of Kearney and Trecker Corporation.)

part cannot pass through until a release switch is pushed. In this hold position (Fig. 4-16), the part waits for an operator to arrive and complete the unclamping and reclamping. A switch is then pushed and the part is free to move through.

Another method for part stress relief is to route the part back to the load/unload area. This eliminates the necessity of bringing the operator to the part. Also, the tools required to unclamp or reclamp the parts are readily available in the load area since they must be used to load and unload the parts. The disadvantage of bringing the part back to the load/unload area is the time delay between successive operations. This time delay will require more work in process to maintain machine utilization, thereby increasing the complexity of the system operation.

4.6.4 Quality Monitoring and Adjustments

Yet another type of manual station is for quality monitoring and adjustments. These are not necessarily separate stations. Usually this monitoring is done while the part is on the work table of the machine. In this situation, the operation will start in the machine and might make a trial cut of the part. Then, operation will be suspended (freedhold) until an operator switches back to automatic operation. During this time, the operator can measure the accuracy of the trial cut and make offset adjustments to the machine. The advantage of this type of monitoring is that the chance for detection and correction are available before the part is actually machined. The disadvantage is that automatic operation is suspended until an operator responds. At times, this will increase the operation cycle and provide an opportunity to lose control over the cycle time.

Overall, many diverse, manually controlled stations are interspersed among automatic equipment. As far as the FMS is concerned, this can be compatible as long as control over the cycle time is maintained; that is, automatic operation is suspended until some form of manual intervention is completed and automatic control resumed only when the operator decides. Allowances are made in the design but if the actual times become longer or operator response time is not predictable, the FMS operation will not reach stationary features. This will cause what is commonly referred to as "surging" in the system's operation: the system will seem to pulse in its level of activity. At some point, demand to resources will be uniform; at other times, demand for a resource might result in a large queue. Once this queue develops, a long recovery time is usually required before stationary operation resumes. The most frequent cause of this surging is the breakdown of a machine but uncontrolled manual operations are the second leading cause. Part III contains a discussion on design and fault recovery allowances.

4.7 IN-PROCESS STORAGE FACILITIES

When FMS was first introduced in the early 1970s, the need for in-process storage within the system was included only for component failures or storage of empty pallets and fixtures. It was felt that with this level of integration and automation, the just-in-time characteristic would be a typical feature of FMS. In other words, no storage was necessary since future activities could be predicted and all motion could be anticipated.

The assumption that all activities could be predicted turned out to be false, and thus the entire concept of just-in-time material movement never became a reality. Therefore, to account for this lack of perfect information, some buffering was needed between the handling system and machines. This buffering of parts is called in-process storage. Within the FMS, there are three types of in-process storage facilities, a dedicated storage and retrieval system, dedicated equipment and station work table storage.

4.7.1 Dedicated Storage and Retrieval System

The first type of in-process storage is a dedicated storage and retrieval system where parts/pallets can be moved for temporary storage between operations. This type of storage is located away from

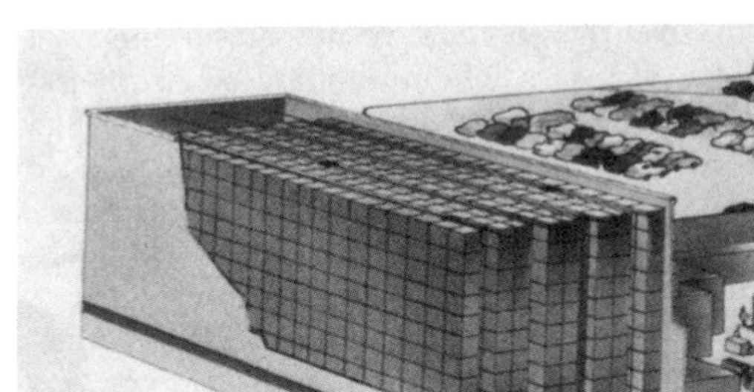

FIGURE 4-17 Automatic storage and retrieval system.

any station; consequently, a material handling system is required to pick up and deliver all pallets to this area. It is possible for this store to be centralized to the entire factory or local for the FMS. In any case, the location of the store will depend upon the response time necessary to interface with the machines.

The physical equipment of the central store can be as elaborate as the automatic storage/retrieval system (AS/RS) shown in Fig. 4-17, where all pallets are delivered and picked up from a common area or a row of pallet stands, each with direct access to the material handling system. No matter what the equipment makeup, the purpose of the central store is to provide a common storage facility for any part, regardless of which operation it requires next. Storing parts based upon their next required operation is a strategy used in another type of storage facility.

4.7.2 Dedicated Equipment

Like the central store, the dedicated equipment storage facility also provides a common storage facility for parts, but instead of being separately located from all stations, each station has its own small storage area (Fig. 4-18). Parts can be delivered to the station and when the station is available, the part can be pulled into the work table without the need of the material handling system.

The physical equipment for this type of store can be pallet stands, rotary pallet changer or multiple part carousel. When pallet stands are used, one will be designated for delivery and another will be designated for pick up.

In this type of storage, parts will only be delivered to those stations which can perform its next operation. Parts will not be free to move to any available position. Some enhanced controls,

FIGURE 4-18 In-process storage facility with direct access to a station work table. Stands can be used for incoming or outgoing work. (Courtesy of Kearney and Trecker Corporation.)

which are discussed in Chapter 5, are available which remove this restriction of where a part can be stored.

4.7.3 Station Work Table Storage

A third type of in-process storage utilizes the station work table to temporarily store parts. No FMS designer will admit to using an expensive machine tool for storage but in operation many times it ends up this way. In this type of in-process storage, the part which just completed an operation remains in the station until a position exists where it can be moved.

4.8 MATERIAL HANDLING SYSTEMS

The material handling system within the FMS has the responsibility of picking up and delivering pallets to all stations within the system.

4.8.1 Nonaddressable Material Handling System

Two classes of systems exist, nonaddressable and addressable. Nonaddressable material handling systems include free conveyors, synchronous and nonsynchronous conveyors.

Free Conveyor

The free conveyor is a continuously moving chain or belt where pallets and parts can be placed and moved to their next station. Figure 4-19 shows a circular conveyor with all machines located around the outside. All parts move in the same direction and at the same speed so that collision avoidance is only needed when a pallet enters or departs the conveyor. Another characteristic of the free conveyor is that parts can pass one another since the material handling system is separate from the work tables. This feature distinguishes the free conveyor from the other two types of nonaddressable material handling systems.

Synchronous Conveyor

The synchronous conveyor is most frequently found in a synchronized transfer line. In this type of material handling system, all parts and pallets move at the same speed and at the same time but they cannot pass one another. This characteristic is found in applications where all parts require identical operations sequences. It is possible for these operation times to be balanced (made approximately equal in duration) for all stations in the line. In this application, when the longest cycle operation is completed, all parts

FIGURE 4-19 Circular conveyor with switch mechanism at both ends. Fixtured parts and pallets move directly along the conveyor. (Courtesy of Ingersoll Rand—Jerry Cahill.)

advance to the next position in the conveyor. These positions are usually always station tables and thus no in-process storage facilities exist other than the table itself. When the series of operations cannot be balanced, then some buffering is needed. This is the main characteristic of the nonsynchronous conveyor.

Nonsynchronous Conveyor

The nonsynchronous conveyor will use station tables as part of the handling system but in-process storage positions will exist between stations. These positions are in the conveyor and the number of pallets and parts which can be stored is determined by the physical size of the pallets and the distance between the work tables. This buffering of parts between stations alleviates the inefficiencies of different cycle times. This allows for two features which are important to the FMS.

First, because of in-process storage, the nonsynchronous conveyor allows different part types to be run concurrently in the system. This is an important characteristic of FMS. The second feature is the ability to design operations which do not have to match to some overall cycle. Thus, the number and type of stations can be accurately determined without the need for extra stations to account for decreasing the cycle. For these two reasons, the nonsynchronous conveyor has much better application in the FMS.

As the need for part movement flexibility (which is independent of the variety of parts currently in the system) and operation durations increase, the ability to move a pallet from point to point becomes a necessity. This type of movement is the primary difference between nonaddressable and addressable type material handling systems.

4.8.2 Addressable Material Handling Systems

The addressable system consists of a physical vehicle which is shared by all parts and pallets, moves no more than two parts simultaneously and can move any pallet directly between any two stations. The advantages of the addressable system are its direct control over material flow and its ability to move parts and pallets based upon station availability. There are many different types of addressable material handling systems.

Towline

One of the first to be applied to the FMS is the towline. The towline consists of a cable or chain which is located in the floor and is constantly moving. A slot in the floor allows the vehicle access to the cable and also provides a guidance path (Fig. 4-20). When the vehicle is required to move, it latches onto the moving cable, which pulls it along. When it is required to stop, the vehicle unlatches and coasts to a stop. Stopping bars will come out of the floor to provide accurate positioning for the vehicle. Many individual vehicles can be moved via the tow line and collisions avoided by use of "zones."

The line layout is comprised of zones, which are designated areas within the entire system. Often a simple control strategy is used which prohibits more than one vehicle from occupying any zone at a time. As long as this one rule is observed, vehicles are free to move independently of any others. With these two characteristics, the motion in the tow line is unidirectional which eliminates deadlock situations.

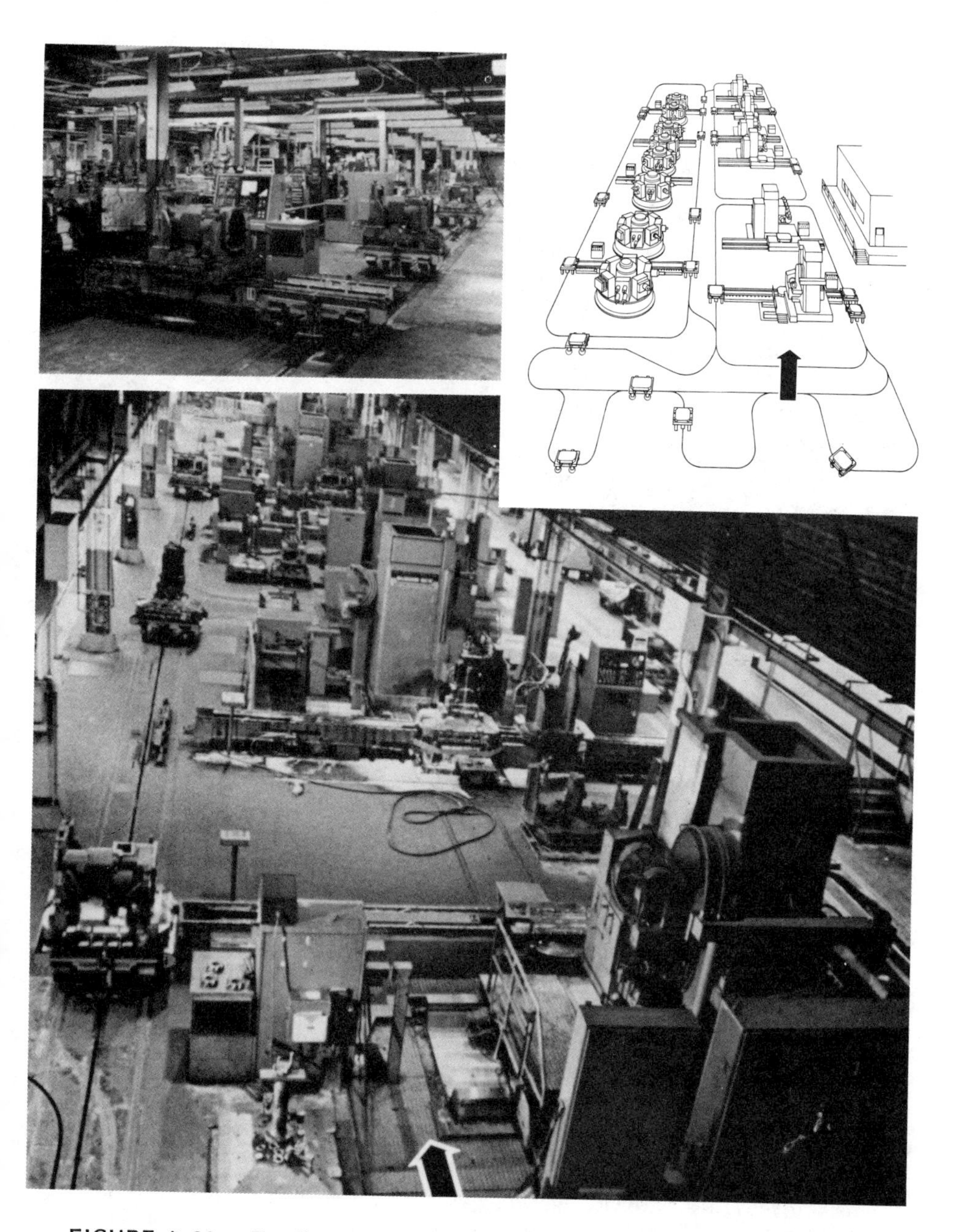

FIGURE 4-20 Towline. Tow chain is embedded in the floor with a mechanism for individual cart movement. (Courtesy of Kearney and Trecker Corporation.)

FIGURE 4-21 Rail guided vehicle, which moves bidirectionally along the rail to service all stations in a line. (Courtesy of Detroit Diesel Allison—Steve Fox.)

Bidirectional Vehicle

One addressable material handling system which is usually bidirectional is the rail guide vehicle (Fig. 4-21). The rail guide vehicle runs on a railroad type track and can be powered by a reversible chain or rack and pinion. With the chain type drive, each vehicle is connected to its own chain for independent motion. Bidirectional motion is obtained by reversing the chain. The rack and pinion drive vehicle also runs along a rail but a gear running in a rack provides the motion. In order to change directions, the gear or pinion is reversed.

One major concern in both types of rail guide vehicles is how to avoid collisions when multiple vehicles run on the same rail. Two strategies are used. One of the simplest is to prohibit both vehicles from moving at the same time. That is, while one is loading or unloading a pallet, the other is free to move. No other vehicle can begin movement until the active one stops. A second strategy is to restrict the area of the FMS in which the vehicle can serve. Thus, one vehicle might service the left to the central area and the other might service the central to the right. As more vehicles are added, collision is avoided by isolation.

It is not always possible to restrict motion of the vehicles and maintain the flexibility of addressable over nonaddressable type materials handling systems. In many applications, overlap zones are permitted where both vehicles can enter but only one can remain at a time. Because each vehicle must advance from its original direction into the common zone, deadlocks are avoided. This permits parts to access a greater portion of the FMS. The control strategy of this overlap zone is similar to that for controlling vehicles in a towline system.

Isolated Vehicle

Whenever multiple vehicles are added to a rail guide system, the transporter capacity does not increase directly by the number of vehicles. This is best described in terms of a two-vehicle system where the carts are isolated into separate areas with a common area in the middle.

Two strategies are available for design of an FMS with an isolated vehicle system. First, stations of the same type can be grouped together and when a part requires an operation from one of these stations, the part is transferred by the vehicle which services this area. The second strategy is to separate the stations so that both sides have identical station makeups.

The advantage of grouping like stations is that backup is available when one station is down. The disadvantage is that some means must exist for the parts to be transferred from one end of the system to the other so that access can be achieved to all stations. However, now one move is required to bring the part back to the shared area and another move to take it to its next station; thus this transfer becomes an operation penalty.

The second strategy eliminates the need for transport transfer by requiring a station of each type to be located in both ends of the FMS. No matter which way the part moves, it can have all of its operations performed before returning to the common area. Unfortunately, such a layout does not provide for flexible backup of downed stations. For example, suppose a lathe machining center was required in both ends of the system, but two were sufficient for the entire systems needs. Therefore, one machining center would be placed in each end without either backup or transport transfer facilities. Another negative feature is that additional station capacity might need to be available to make sure that each end (or subsystem FMS) has balanced requirements so that efficient ultilization of the stations can be achieved. With the variety of part types, some might require long operations on one station and others might require different demands of the stations. With the dynamic operation of FMS, it is impossible to predict how efficient utilization of the equipment will be maintained.

In this second strategy, efficient use of the transport system can be achieved but at the risk of inefficient station utilization and restricted backup support of failed equipment. The trade-offs between transport efficiency and station efficiency are a common decision variable in FMS. Part III contains a detailed discussion of the trade-offs of component efficiency when integration is required.

Overhead Crane

Another vehicle system which is similar to the bidirectional vehicle is the overhead gantry crane. The crane is guided by I-beams which run above the stations The gantry, located over the delivery and pickup points at the stations, has an arm which can latch onto the part and pallet. When the pallet is latched, it is raised to a travel position, clearing any other pallets which might be located in its path.

Just as in the case of the bidirectional vehicle, multiple cranes can be placed in the same gantry but collision avoidance must be accounted for. Also, similar inefficiencies occur from two bidirectional vehicles in the same track.

Automatic Wire Guided Vehicle (AGVS)

For the reasons mentioned above, if there is any doubt that more than a single vehicle is required to service all stations in the FMS, an individual cart system is used, such as the tow line or the automatic wire guided vehicle system (AGVS).

The AGVS contains any number of vehicles which are controlled by radio signals, transmitted through wire embedded in the floor. Each vehicle (Fig. 4-22) has a receiver which can detect the radio

FIGURE 4-22 Automatic guided vehicle. The AGV follows a wire embedded in the floor and is positioned at the station by use of locating switches in the floor or attached to the station. (Courtesy of Ingersoll Milling Company.)

signals and be individually moved by frequency through zones. These zones have the same representation as the zones in the towline description. No more than one vehicle can occupy a zone and once a vehicle is observed in a zone, no other vehicle can enter it. The operation of the AGVS is similar to that of the towline but does have some advantages.

The first advantage is the ease and flexibility of installation and layout. Installation of the AGVS is accomplished by cutting a groove in the floor and laying a wire covered by a protective sealant. The major concern is that the wire does not violate the minimum turning radius of the vehicle, so that they will not depart the wire and lose the radio signal. The receiver can detect the signal about three inches from both sides of the wire but once it is any farther away, the vehicle will not be able to receive directions.

Since an AGVS travels over the shop floor guided by a wire, accurate positioning is not guaranteed as in the case of the rack and pinion vehicle. To resolve this problem, the vehicle can be aided by positioning cones which are permanently attached to the floor. As the vehicle stops, its support pins are lowered, covering the cones. As these pins come into contact with the cones, accurate positioning is achieved. In most cases of pallet transfer, instead of lifting the entire vehicles, only the transport table need be lifted for alignment with station tables or in-process storage facilities.

The sacrifice for this freedom of layout and movement is in the power supply. Each AGVS must contain its own power source for all movement and lifting of the transport table. This power is provided by a battery which usually has the capacity to maintain eight hours of service. Many methods are available to extend this eight hour power supply by either intermittent charging of the battery or use of some auxiliary power for lifting the transport table. The auxiliary power is obtained through use of the power post. Vehicles stopped to load or unload pallets can plug into an auxiliary power source at these posts. This power can be used to raise and lower the transport table, which is the largest energy demand, and to charge the battery while the pallet transfer is taking place. But no matter how much auxiliary power is available, the batteries will eventually need to be replaced. When the power level of the battery drops below a certain point (usually enough power remains to complete a couple more assignments and be able to safely return to the battery changing area), signals are sent to the vehicle control which will direct the vehicle to the replacement area. Once it arrives in this area, either an operator will replace the old battery with a charged one or some mechanical means will be available for automatic battery replacement. Very seldom is the vehicle taken out of service while the battery is charged since the cost of the additional battery is much less than the cost of an additional vehicle. The replacement usually takes less than ten minutes and the used battery is placed in the charging area.

Again with the AGVS, trade-offs between flexibility and efficiency must be evaluated. The AGVS is by far the most flexible in terms of FMS layout and the ability for extension or change after installation. This is gained by relinquishing some efficiency in operation in the limited power supply. Additional time is also required for accurate positioning of the transport table. The AGVS has become almost the "typical" material handling system for FMS but the trade-off cannot be ignored when considering the best material handling system. Part III contains a detailed discussion of material handling evaluation for FMS.

Because of its flexible layout, it is possible for the AGVS to be equipped with sidings where it can travel when it needs to service a station. Each station will contain a siding and vehicles will be free to pass other vehicles which are servicing machines. This is not as common in the towline. There, vehicles usually stop in the main line to service stations and until they move ahead, any vehicle coming from behind must wait until the service is complete. Sidings can increase transport capacity by as much as 20%. Unfortunately, additional floor space is required by the machines. Instead of being two vehicle widths apart, they must now be four vehicle widths apart or over twice as much floor space will be needed for the transport system. One development which attempts to overcome this problem is the "wireless vehicle."

Wireless Vehicle

The wireless vehicle is guided by a signal transmitted from its destination. This permits vehicles to take a direct path from their current position to any destination without interference from other carts. Since no specific path is required, the vehicle is free to move around any cart which is stopped to load or unload.

The trade-off of this flexibility is the increase in control complexity. Now the control must have additional sensors and must be aware of the position of all vehicles in order to maintain effective collision avoidance. Because of the freedom of movement, the number of potential collision situations increases. There will also be situations from which the control will not be able to automatically recover. Some may result in deadlocks and require manual intervention to restore system components to automatic operation.

4.9 CONCLUSION

The hardware components in an FMS are as diverse as the equipment for the factory. Because of this variety of components, any manufacturing system which integrates several of these components might be called an FMS. However, the definition of FMS must extend beyond the hardware and include the computer control hierarchy. This computer control is the channel which integrates (coordinates) the activity. Hardware integration is achieved by uniform material handling and aligning vehicles with pallet stands, but these are useless without the controls of each component being able to confirm status for compatible operation.

5

Control Components of Flexible Manufacturing Systems

Many FMS definitions include only the hardware description of the system and ignore the control components. But because the computer control integrates and coordinates the activity of the hardware components, it is this which differentiates FMS from all other types of factory automation.

As described in Chapter 4, each hardware component within an FMS can be some type of island automation with its own computer control. When these islands are brought together, the computer controls can no longer operate as an independent computer with independent information. Rather, information must be passed from one computer to another, forming a bridge within the FMS concept. Although there is little question of the importance of the computer control within the FMS, manufacturing managers usually agree that it is the weakest link in the system.

One primary reason for the weakness in the computer control is that each component in the FMS has not been specially designed for FMS but has been installed as independent machinery. Therefore, when this equipment is installed, much more emphasis is placed upon the hardware integration. For example, many hours of engineering are allocated to provide a uniform material handling means, whereas relatively few hours are devoted to computer integration. This strategy of integration allows the hardware to act as one system while preserving the control characteristics of each component as if they were characteristics of each component are preserved as if they were operating as an island. The central computer must then coordinate all computers to work together because no other component has the expanded control necessary to carry out these integration decisions.

For example, the control for the machine center can perform sequences of tool changes and cuts, enabling it to complete an operation automatically. However, in the FMS the part must be properly referenced or positioned to the machine. Instead of expanding the control to incorporate reference decisions, these accuracy decisions are made in less direct methods. The accuracy decision is approached from two means. First, repeatability of the hardware is required. This means that all machining centers are as identical as possible and that all fixtures are within designed tolerances. But because designed tolerances can never be completely maintained, predetermined offset calculations are made to account for the lack of repeatability between fixtures and are passed on to the machine control. The control then uses these offsets to compensate and bring fixture machine reference to within tolerances.

The proper integration technique would extend the machine control to repeatedly reference all parts within design tolerance. This would be accomplished by requiring the control to "locate" the part and adjust offsets by using sensor feedback. In this case, repeatability in the hardware would not be necessary and any offsets could be dynamically determined, rather than using results from measurements taken in a controlled situation.

Another example of the lack of integration software is in the area of vehicular traffic control. The typical automated material handling system is capable of accepting three types of commands: (a) move, (b) move to pick up or (c) move to deposit. With these three commands, a destination is provided, the shortest path is found and the vehicle moves along this path. When another vehicle blocks its path, it remains until the blocking vehicle moves out of its way, avoiding collision. Vehicle movements must be coordinated with station availability. If the coordination is not present, deadlocks can occur between two vehicles with no automatic method for recovery.

One method for reducing the impact of vehicle interferences is to construct a layout in which a free path will exist between any two points. This is the advantage of the AGVS over other types of handling systems. This solution, however, requires the addition of more hardware to overcome the inadequacies of the control. The proper FMS application should not require additional hardware but the integration decision should be made at the level where it has the greatest impact.

This chapter contains the description of the control of each hardware component. As in Chapter 4, this presentation is not intended to review how the control currently works. Rather, it focuses on integrating information passed between various hardware components. These components are condensed into three major areas: stations, transport and storage. The control computer for each area and the expansion necessary to include integration type decisions are described in the following sections.

5.1 MACHINE CONTROL

Computer Numerical Control (CNC) is the typical control for the machine within the FMS. The CNC contains a central processing unit with memory and an operator panel (Fig. 5-1). These computers contain a specialized operating system which is designed specifically for interpreting program instructions and activating appropriate switches.

5.1.1 Servo Motor Feedback Control

The CNC is designed to accurately position the axes of the machine tool. This is accomplished by using resolver type motors with inductance feedback (Fig. 5-2). As the servo-motor functions, the number of revolutions and positions in the operation is constantly reported to the CNC. Distance moved is directly related to the number of revolutions of the servo motor. Therefore, with appropriate scaling, the CNC can regulate distance by monitoring the servo-motor feedback. Most positioning can be done in two speeds: rapid movement of over 200 inches per minute or creep for accurate positioning of less than ten inches per minute. As long as the calibration of the servo-motor to distance is accurate, the CNC can position in any axis with high accuracy.

The ability to control servo-motors and monitor position feedback is one of the unique features of a CNC. The other features of the CNC are very much like the microcomputer. The CNC will contain memory where the operating system, more commonly referred to as the executive, and application programs, also known as part programs, can be stored and executed. Some CNC provide bubble type memory which permits the information to be retained even if power is disconnected.

5.1.2 Part Program Storage

The executive program contains information pertaining to the calibration of each servo and a set of instructions which permits application programs to maintain direct control over these motors.

Part programs similar to the example shown in Fig. 5-3 are written in a machine language specifically for the executive program. A part program contains a set of instructions to carry out a sequence of cuts and tool changes to complete an operation on a single part. Several separate part programs are usually stored in the memory and

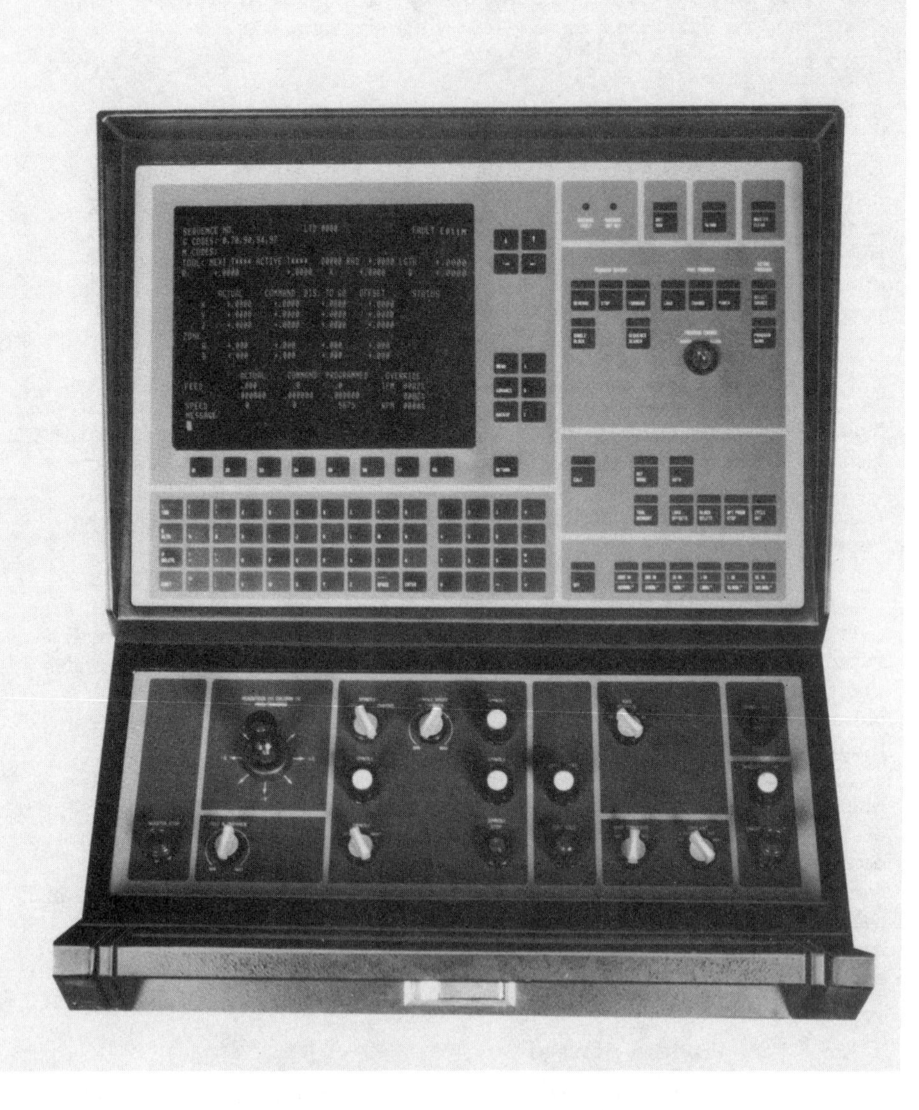

FIGURE 5-1 Operator panel.

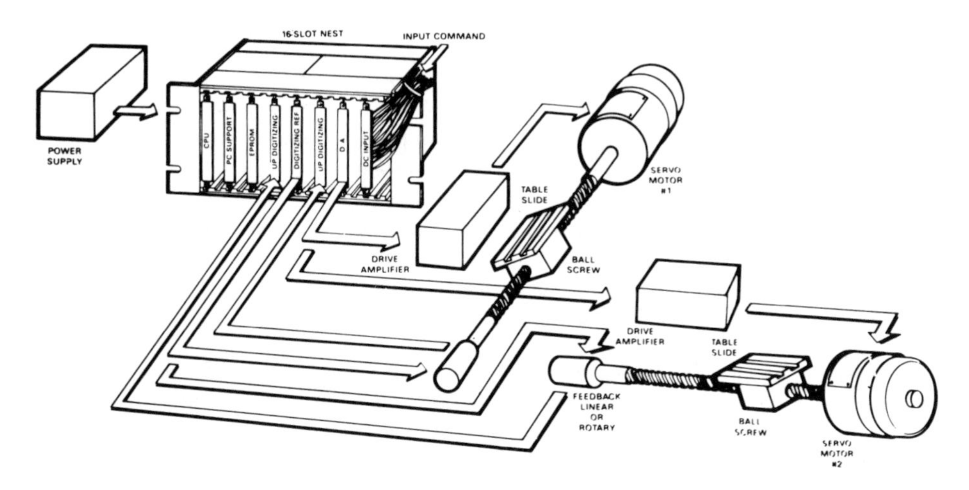

FIGURE 5-2 Servo motor.

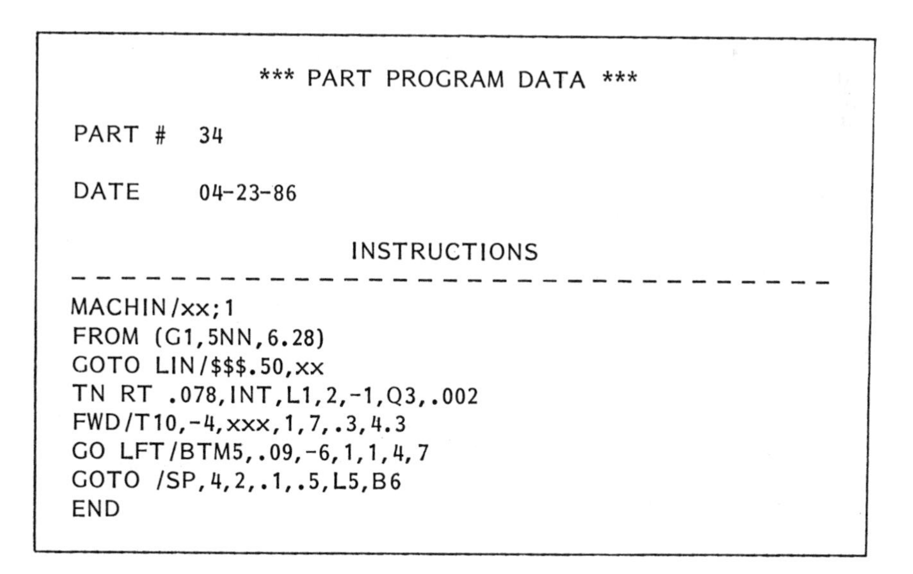

*** PART PROGRAM DATA ***

PART # 34

DATE 04-23-86

INSTRUCTIONS

```
MACHIN/xx;1
FROM (G1,5NN,6.28)
GOTO LIN/$$$.50,xx
TN RT .078,INT,L1,2,-1,Q3,.002
FWD/T10,-4,xxx,1,7,.3,4.3
GO LFT/BTM5,.09,-6,1,1,4,7
GOTO /SP,4,2,.1,.5,L5,B6
END
```

FIGURE 5-3 Part program.

any one can be activated. However, the CNC is a single processing system that allows only one program to be operational at a time.

The CNC for the FMS has been expanded to include the ability to communicate with the integration computer. This computer contains part programs for all stations which are transferred to and from any CNC. Often, the CNC can send and receive information while it is running a part program, providing status information at regular intervals and accepting instructions for future activities. This ability to communicate while running a part program requires the CNC to be able to perform multiprocessing functions which are essential for the complete integration of the machine to the rest of the system.

Another important component of the CNC is the ability to store part programs. This is especially beneficial for the FMS application, since a variety of part types can have operations performed at the same machine.

There are two ways to manage part programs within the FMS.

5.1.3 Part Program Management

In the first method of part program management, the CNC stores only the active part program. While the part is being transferred to the work table, the appropriate part program is "downloaded" to the CNC, replacing the previous one (Fig. 5-4). The benefits of this

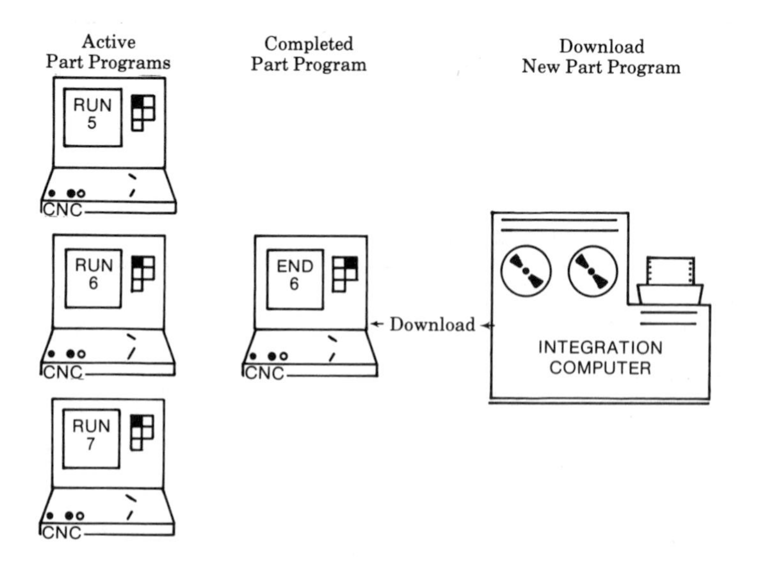

FIGURE 5-4 Part program download.

approach are that as part programs are changed, it is assured that these changes are immediately used and original copies do not remain in any CNC. There is also less chance of selecting the wrong part program when only one is stored in the CNC at a time. The disadvantages are that a "download" is required each time a part is moved into the work table and this communication might take longer than the actual part transfer. Portions of the part program could also be altered because of noise in the communication increasing the potential for errors in the operation. But when multiple part programs are stored in CNC memory, this problem of part program accuracy is reduced.

In the alternative method for part program management, multiple part programs are stored simultaneously in the CNC. In this approach, when the next part is selected for machining, a review is made to ensure that the appropriate part program is stored in memory. If so, it is activated once the part arrives in the work table. If it is not already in storage, a "download" must take place. The advantages of not having to download are that the chances of receiving a corrupted part program because of communication errors are reduced and delays due to timing inconsistencies are eliminated. The disadvantage is that each CNC might contain a copy of the part program and if any alteration is made, all copies must be updated. But this can be accomplished with the use of part program management software. The program at each machine is recorded, and when needed, the CNC version number is compared to the latest version and a download will occur only when these versions do not match. The ability to store multiple part programs has advantage over single part program storage except when uploading from the CNC is required.

In uploading (Fig. 5-5) the part program in CNC is transferred to an integration computer. This feature is useful when testing a new part program for accuracy. Modifications to new part programs can also be made using the edit features of the CNC. Once these modifications have been made and checked, it is beneficial to be able to send this program version back to the integration computer. If upload capabilities are not available, the edits must be reentered without testing to make a new tape, which starts the debugging process all over again. However, when uploading in a system with multiple part program storage in the CNC, the possibility of replacing revised part programs with old versions is increased. This can be reduced through proper procedures and program securities. But when many versions reside in the same network, it is possible for an older version to replace the current one.

Features such as servo-motor feedback control, part-programming, transfer and part-program storage are fundamental in the operation of any CNC. There are several additional features for the

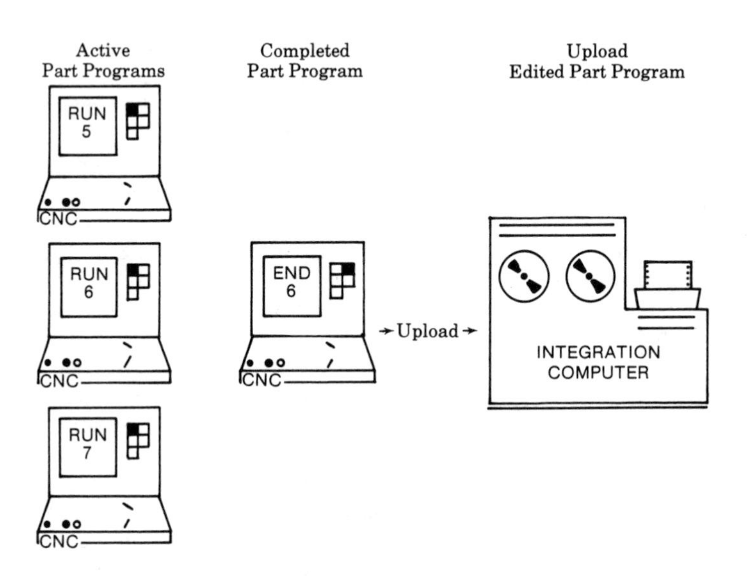

FIGURE 5-5 Part program upload.

CNC which are necessary for an FMS application. These include detailed tool monitoring, extensive fault detection and recovery, and automatic adjustment for quality control.

5.1.4 Detailed Tool Monitoring

One functional area where the CNC in an FMS must differ from the CNC as island installation is the monitoring of tool status. An island CNC requires an operator to cycle start and monitor tools while the program is running. In the FMS, no operator is necessary because computer control assumes these duties.

Tool life monitoring is a process which records the amount of wear each tool receives. Figure 5-6 shows an example of a tool life record. This life can be determined in two ways. Each time the part program runs, precalculated wear times can be accumulated to the tool life. These static times are usually recorded from the postprocessor output of the part program and take on constant value. The main problem with this is that as modifications to the part program are made, these constants must be maintained as well or they will not be accumulating proper wear. Another problem is that these times do not account for the time the tool is actually in contact with the part and the effects of feed rates. Since tool wear monitoring involves dynamic data, it is much more accurately

*** TOOL LIFE RECORD ***

TOOL ID	TOTAL LIFE	WARNING POINT	CURRENT WEAR	TIME OF LAST REPLACEMENT
1" DRILL	400MIN	370MIN	200MIN	10-MAR-86:10:05a
1" TAP	600MIN	570MIN	60MIN	15-MAR-86:11:15a
1 1/2" TAP	500MIN	470MIN	400MIN	11-MAR-86: 1:04p
1 1/2" DRILL	450MIN	375MIN	150MIN	14-MAR-86: 3:32p

FIGURE 5-6 Tool life record.

collected by actual usage than by precalculated wear times. For example, a part program may use a 9 mm. drill for five minutes but an operator-altered decision to the part program reduces that time to two and one-half minutes. If static data is used, the corresponding tool usage item must be updated or the recorded wear will be twice as long as the actual wear.

However, actual usage can be collected as part of the CNC executive program. The time at which the drill is loaded and unloaded into the spindle can be logged and this dynamically collected duration will represent tool wear. Sensors can also be used to determine when the tool is actually in contact with the part so that wear might be accumulated for only that portion of time. The advantage of this type of wear accumulation is that it automatically adjusts to part program alterations and some degree of accuracy is ensured.

With the static wear calculation, the energy required to maintain accurate data is often greater than the energy required to routinely check each tool visually, as was done in CNC island installations. When this occurs, an advantage of FMS over conventional CNC is lost and the benefits are reduced. Part IV contains discussion of the benefits and trade-offs of FMS.

Another component of the tool monitoring system is the warning point. When the total wear reaches this value, a signal is given to indicate that this tool is reaching its maximum wear level. The value of this point should be set to provide sufficient time for review and replacement of the tool before it reaches its maximum wear. In some cases, this might be only one remaining operation and for others it might allow as much as one hour for response. In any case, the time which is set must provide sufficient response time to replace tools before they reach their maximum wear.

5.1.5 Fault Detection and Recovery

Another area where the CNC for FMS must be enhanced is in fault detection and recovery. The traditional CNC has the ability to detect many faults such as end of travel for any axis, hydraulic pressure and other damage prevention tests (Fig. 5-7). Once any of these faults are detected, the CNC issues an emergency stop and all activity is suspended until the fault is cleared. Clearing the fault might require simply cleaning chips out of the way and then restarting the CNC with cycle start, or may be as extensive as replacing a component of the machine. However, whenever any fault is detected, the current activity is suspended until an operator intervenes.

Not only is the operator needed to recover from faults, but the operator of a conventional CNC provides a constant check on operations and can often prevent major machine wrecks. In an FMS,

FAULT CODE	DESCRIPTION
01	END OF "X" TRAVEL
02	END OF "Y" TRAVEL
03	END OF "Z" TRAVEL
04	EMERGENCY STOP BUTTON DEPRESSED
05	TABLE ROTATION EXCEEDED
06	TOOL CHANGER
07	TOOL NOT CLAMPED INTO SPINDLE
08	TOOL LENGTH

FIGURE 5-7 List of faults.

there are no operators to monitor the machines; consequently, wrecks which could have been avoided are not. For this reason, sensors are added to fill in this vital monitoring system.

One such monitoring technique is horsepower monitoring. Horsepower is a measure of the amount of work being done. It is possible that each cut may have some predetermined horsepower level associated with it. The monitoring system can feedback the actual horsepower to the CNC and this can be used to compare to expected levels. If the actual horsepower is outside control limits, this will result in a machine fault. Operator intervention is still required but it is hoped that major wrecks of the machine can be avoided.

Another sensor for early warning of a problem is vibration detection. If the amount of vibration exceeds a prescribed limit, the CNC can suspend its current action in expectation that major problems have been avoided. The operator, who is required to respond, can restart the machine in a controlled mode and attempt to diagnose the problem without further damage to the machine.

In all sensor applications within the CNC, each is intended to monitor an operational characteristic and generate a fault if any one characteristic is detected to be out of control. As a result of any fault, operation is suspended until an operator responds, diagnoses the situation and establishes a recovery procedure. To reach higher degrees of automation, some form of automatic recovery from these faults is necessary. For example, when a tool is about to break, the horsepower will increase. If this could be detected in time for the CNC to suspend the cut and retract the tool, the length could be measured. If it was determined that the tool was not of appropriate length, an automatic recovery procedure could be implemented to replace it with a backup tool. Once the replacement had been completed, the cut could be restarted from where it was suspended without the need for operator intervention.

Sensor information can be used to provide the capability for automatic recovery from faults, but more importantly, it has the ability to provide an early warning signal. Without it, a problem can progress too far to utilize a single type of recovery. For example, if a tool breaks, part and machine damages must be assessed before replacing the tool. Because of this additional process, it is impossible to establish a predefined recovery procedure.

5.1.6 Automated Adjustments for Offset

Another area in great need of automation is the adjustment of offsets. As much accuracy as possible is incorporated into the hardware, but it is impossible to expect that machines and pallets are identical. Therefore, small adjustments, known as offsets, are needed to compensate for differences beteen fixtures, pallets, parts, and tools. These include tool offsets, pallet offsets, and fixture offsets and can vary between machines.

Tool Offsets

Tool offsets (Fig. 5-8) are the programmed adjustment of each tool from some predefined length. These offsets are the difference between the actual tool length and that which is expected or the preset tool length. When the tool is inserted into the tool holder, its exact length is measured and reported by the use of a tool setting station. This length is subtracted from the preset length and when this tool is entered into a tool storage of a machine, its offset must also be entered into the CNC memory. The alternative is to require that all identical tools, no matter which machine, have exactly the same length. The advantage of this is that any tool can be replaced in any machine at any time. The disadvantage is that tooling costs will be higher because tools which have been resharpened several times will not be able to be reused in the system. Although it is possible that these tools could be used on machines outside the system, the chance of reusing tools decreases without the use of offsets.

This reuse of tools is the primary advantage of offsets. The disadvantage is that it introduces chance for inconsistent information with the hardware. Each time a tool is replaced in a machine, its offset must be entered into the control memory. In most cases, this is accomplished by strict procedures of manual tool replacement.

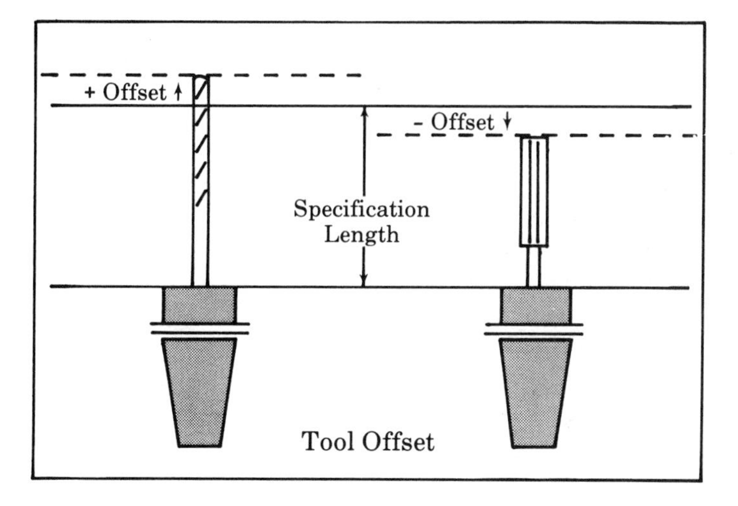

FIGURE 5-8 Tool offset.

The operator identifies the tool offset and when the tool is entered into a machine, the offset is entered directly through the panel of the CNC. But there is always a possibility of operator-entered offset errors. Small errors in offsets (less than 0.005 inch) will usually not result in any fault other than tool breakage. However, if these offsets are in gross error, it is possible to damage the tool changer. These offsets are used to establish depth of cuts and in the case of a milling cutter, further damage can result from taking too heavy a cut and overloading the machine power system.

An automatic information updating system is required whenever a tool is replaced to ensure that offsets are always accurate for the tools which are stored in the machine. The stations which are used to determine tool offsets can now record this value and electronically transmit it to an integration computer. This information can be recorded and automatically read by use of the tool holder identification which was described in Chapter 4. Once this tool is needed by the machine, the CNC will need to request the offset from the supervisory computer. The offset can then be downloaded, guaranteeing accurate information through automatic collection and distribution. This strategy eliminates the need to distinguish between manually and automatically replaced tools.

In automatic delivery and replacement of tools to each machine, it is useless to deliver and update the hardware without automating the information update as well. The only way to automatically replace tools without updating the offsets is to preset all tools to defined lengths, keeping all offsets the same. Proper information handling is necessary for best results in the application of automated tool delivery.

Pallet Offsets

Another type of offset, depicted in Fig. 5-9, is the pallet offset. The offset is used to account for differences between pallets and to bring the pallet to exact reference with the spindle. Often each axis can have its own offset. However, because of the complexity in maintaining three individual offsets, usually there is no allowance for offsets in "x" and "y" and individual pallet characters are only accounted for in the "z" axis. In any case, these offsets are a constant quantity for the life of the pallet and will be used for all machines. Because there is no way to compensate for all possible

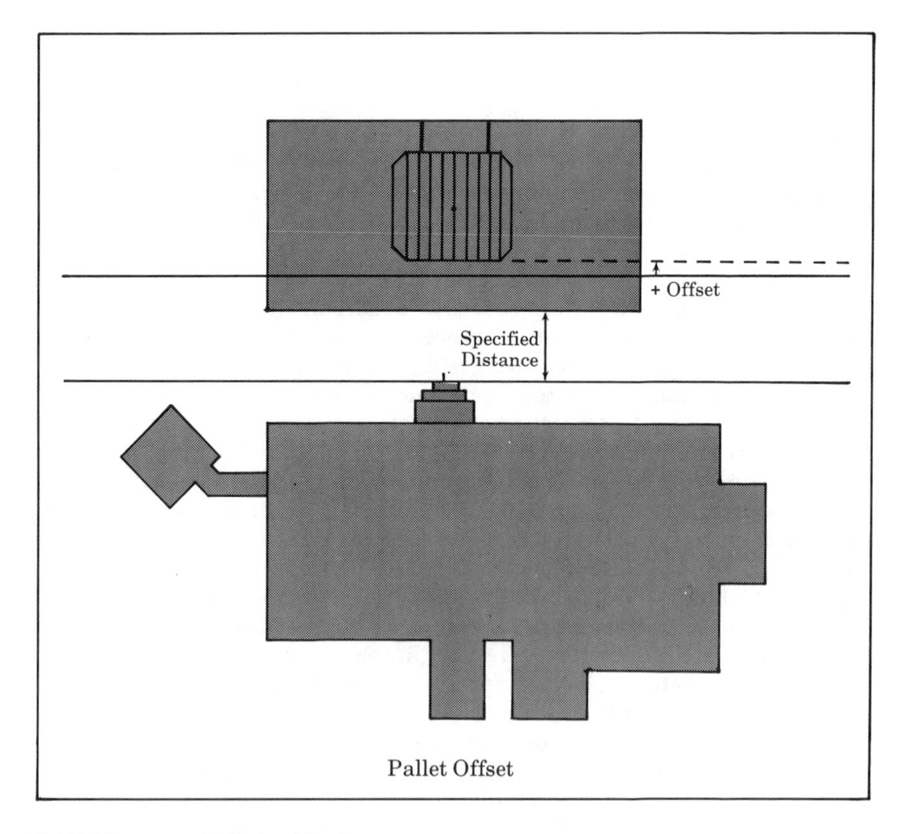

FIGURE 5-9 Pallet offset.

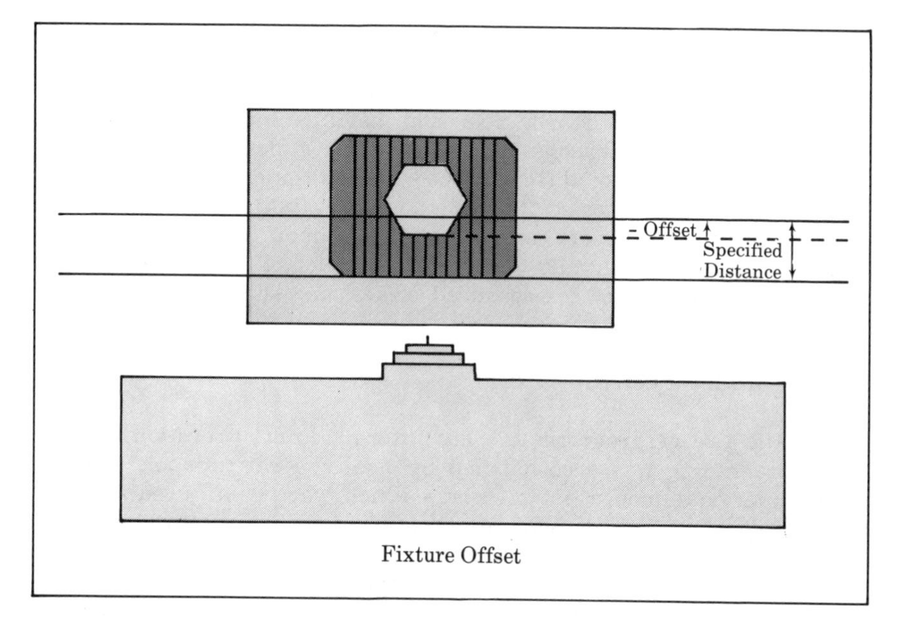

FIGURE 5-10 Fixture offset.

axes-offset combinations, it is assumed that each machine spindle must be perfectly perpendicular to its work table.

In order for several machines to maintain the same reference to a pallet, the spindles and work tables of each machine must reside and operate in the same plane. This is like installing machines which are exact replicas.

Fixture Offsets

The fixture offset, pictured in Fig. 5-10, provides one more degree of freedom in meeting predetermined part to machine reference. This offset measures the differences between various fixtures being placed on the same pallet. For example, a pallet which can be used for multiple fixtures will have a pallet offset for referencing the pallet to the station. Each fixture will then have an additional offset to compensate for the exact position of the fixture within the pallet.

The purpose of the fixture offset is to provide the capability to accurately reference parts to the station without requiring each fixture to reference off of the same point within the pallet. This offset does not provide any greater tolerances in repeatability of the fixture/pallet between stations and is only needed when fixtures on the pallet will be exchanged.

The use of tool, pallet and fixture offsets is for maintaining the quality of parts. Their true role is in establishing repeatability between pallets and stations. The part will "appear" the same to any station with use of different tools, pallets and fixtures. These offsets permit the decomposition of the repeatability problem into discrete parts. However, most offsets only allow single axis adjustment. Chapters 12 and 13 describe the challenge of maintaining part quality with the use of premeasured offsets for multiple axis problems.

5.2 TRANSPORTER CONTROL

Before the use of programmable computers, logical operation of integrated machining was accomplished by use of a relay panel. The relay panel contained a set of relays which were configured to exhibit a logical coordination of integrated components. But because the logic varied from one application to the next, a new physical panel was constructed for each application. Not only were there restrictions of having to construct new panels, but this technology was also limited to "if-then" type decision making. Due to these limitations, the programmable controller became a natural replacement for the large relay panels. The programmable controller, shown in Fig. 5-11, often referred to as the PC (not to be confused with the now-famous IBM pc) has additional features beyond the "if-then" decisions of the relay panel. The PC also has the ability to provide conditional branching, subroutines and memory variables as part of its logic designs.

5.2.1 Ladder Logic

The PC is programmed in a special language called ladder logic. Ladder logic gets its name from the logical view of the program which resembles a ladder (Fig. 5-12). The rungs of the ladder are the "if" condition and the supports are the "then" clause. Drawing a decision tree is very similar to programming in ladder logic.

The program is executed by scanning down the ladder rungs while checking the status of memory variables. These memory variables differ from a general computer in that their value is obtained by direct signals from hardware. For example, if a component comes in contact with a limit switch, a value will appear in a corresponding memory variable in the PC. When the switch is no longer in contact, the value of the memory variable will be zero. As the ladder is scanned, these memory variables which reflect hardware status are checked. The logic of the operation is determined by the specific sequence of memory variable review. Since

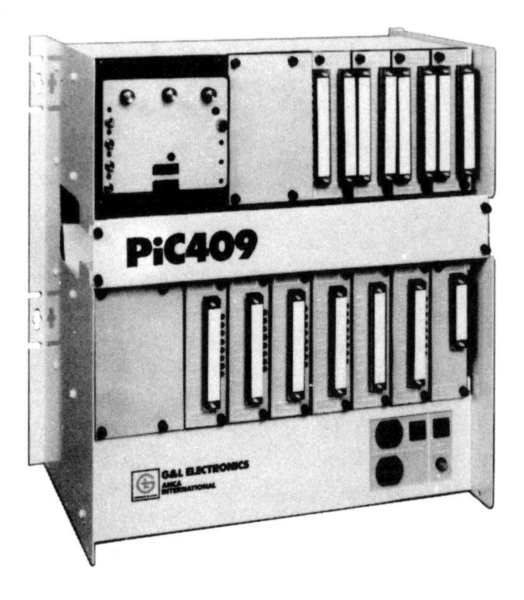

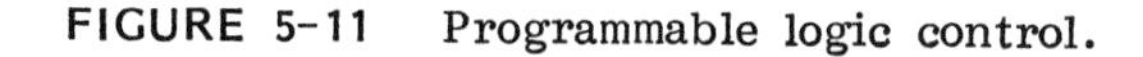

FIGURE 5-11 Programmable logic control.

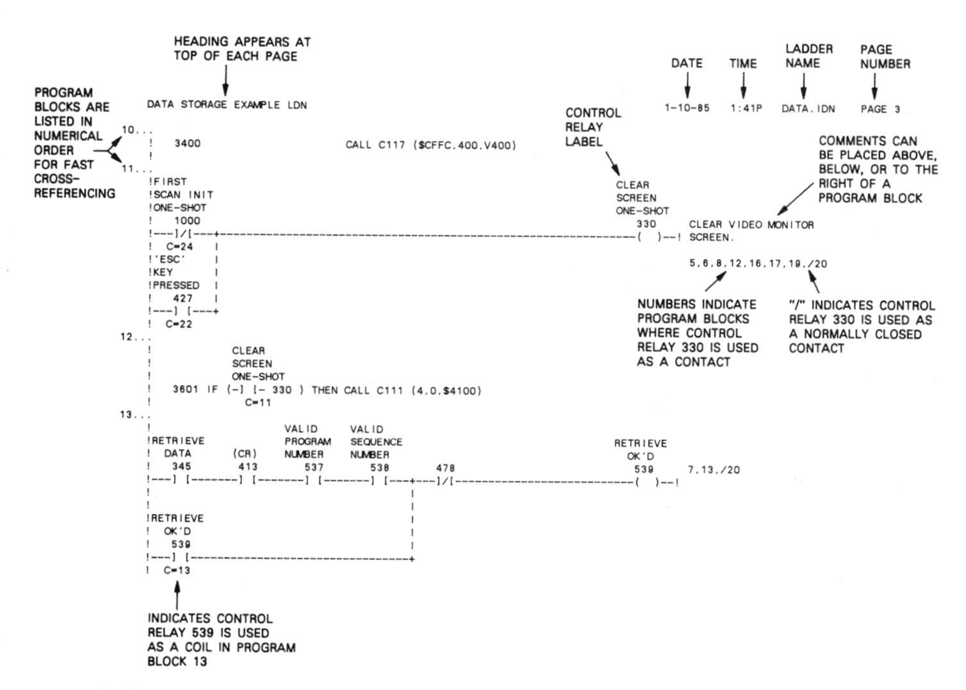

FIGURE 5-12 Ladder logic program listing.

the ladder is scanned from top to bottom, safety checks must be made prior to testing an "if" condition. Failure to do this could result in setting a memory variable status which could damage the hardware.

5.2.2 Interlocks

This sequence of checks, which prevents component damage, is referred to as an interlock. An interlock is not a hardware switch, but a series of safety checks made before any decision for motion is given. Unfortunately, decisions for movement can still result in damage to the machinery if inaccurate information is supplied.

One advantage of the PC is its ability to respond quickly to hardware status change. Because the ladder is scanned hundreds of times each minute, when a switch is made, the resulting motion will appear immediately. Another advantage is in the ease of programming logical motion and interlocks to safeguard hardware from damage.

In the FMS, most decisions are not of the "if-then" type but rather are "on-case" type decisions. Unfortunately, PCs are sometimes unable to handle these complex decisions and they are forced up to a higher level computer. Usually, however, the higher computer is not dedicated to only making decisions for a single component, so there is no possibility of immediate response. To remedy this lack of response, the PC should be able to have direct connection to hardware switches and also be able to make complex decisions. Enter the microcomputer.

5.2.3 Microcomputer-Based Control

The microcomputer (Fig. 5-13), as a controller for the transport system in an FMS, must maintain direct communication with switch status while being programmed in a higher level language. If this combination is achieved, the higher level language will permit complex decision making with direct control over the actions in the hardware. This combination will permit decision making such as traffic control, transporter reassignment and response to dynamic changes in the system.

5.2.4 Vehicle Control

Vehicle control is the set of decisions which most effectively move individual vehicles to carry out assignments. The integration computer will provide requests or assignments to individual vehicles to either move, move and pick up or move and deposit. When multiple vehicles exist, traffic congestion can result. Several methods are

FIGURE 5-13 Microcomputer.

available to relieve this bottleneck, but all require decisions that are not possible in a PC with ladder programming.

Most transport controls provide collision avoidance by not allowing more than one vehicle in any area or zone at a time. Movement is accomplished by pushing vehicles into unoccupied zones and pulling vehicles into zones with departing vehicles. This push–pull type of control utilizes the shortest path by moving vehicles based upon zone availability. This should not be confused with alternative path selection based upon status.

When path selection is needed, alternative paths must be evaluated in terms of all vehicle positions. The information required to make this decision involves all vehicles' current positions and current assignments. Because of this breadth of information, path selection decisions must be made by a microcomputer-based control system. The microcomputer is capable of dynamically determining both paths based upon vehicle status and assignments to find a path with the least interference. Several alternative algorithms are described later in this chapter.

Robots are also controlled with PCs but the language is often more on the order of part program capability rather than ladder logic. Most robot companies have their own robot language, which is a set of instructions to carry out motion. Some of these languages allow for a dynamic response to events, which senses a certain condition and calls upon a subroutine to respond in a preprogrammed manner.

5.3 IN-PROCESS STORAGE CONTROL

The control for in-process storage is as varied as the type of equipment used for temporary storage of parts and pallets in the FMS. When pallet stands are used with each machine center, the control of these stands is usually part of the machine CNC. If pallet stands which are not connected to any machine are used as a dedicated storage area, the transporter PC usually controls the storage. An automatic storage facility will have a PC or microcomputer as its own control. But no matter which combination is used, the control functions identically in each configuration as a means to record pallet position in the system. This tracking system uses readers to detect the identification on each pallet, known as the pallet number. Two examples of pallet recording systems are illustrated in Fig. 5-14.

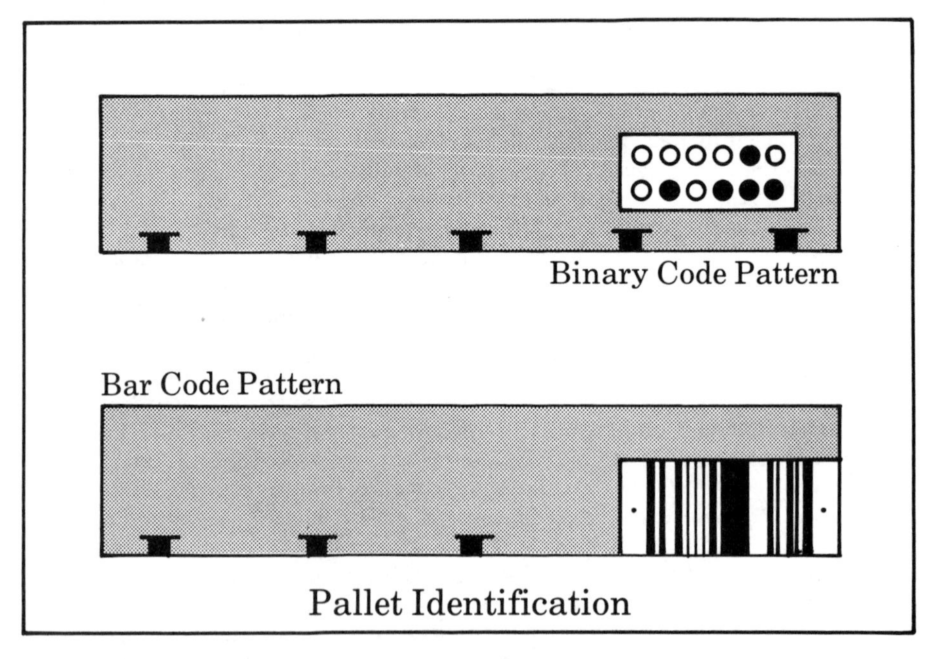

FIGURE 5-14 Pallet ID/binary and bar code pattern.

5.3.1 Pallet Tracking

One specific tracking system uses a series of pins which provide a unique identification of each pallet. A reader containing switches aligns itself with each pin location as the pallet passes over it. As the pins come in contact with the reader switches, the pallet pauses to ensure the reader has an opportunity to detect and receive the signal. This is necessary when the switch does not contain memory and can only send a signal while contact is maintained.

This pause is also used to synchronize the decision information of computer memory with the real position of each pallet. In many control schemes, pallet motion will take place in single steps. These steps might include moving from the work table to a pallet stand, moving to a transporter or departing from a transporter. By giving single moves, the control can monitor the move by using time delays and can determine when an interruption of automatic operation has occurred.

5.3.2 Usage of Pallet ID Switches

When a pallet is commanded to move, the decisive control must know its identity and status. This information must be recorded in the memory of the control. When the motion is completed, the control only needs to know whether the motion is finished and the pallet status can be updated. However, the pallet reader switches are used both to identify pallets and to indicate when a motion has been completed. The former function is needed only at start-up of automatic control and once pallet identity locations are known, they will be recorded as decisions are made. Under automatic control, feedback is only necessary when motions have been completed so that further decisions can be made. In such applications of tracking, the accuracy of the switch is constantly being compared to the reliability of the control's memory. Lack of correspondence between the two usually results in the loss of a pallet and suspension of automatic operation. It might be helpful to use the switches for motion completion only during automatic operation and to monitor their accuracy. In the event that the signal does not match the control memory, one should refrain from suspending automatic operation and send a message that the switch is not relaying expected results. Maintenance can then check the switch. However, it should be remembered that operation of the FMS is not tied to the reliability of each switch in the system. No hardware switch will ever be as reliable as computer memory and should not be required for automatic FMS operation when it is not needed.

FIGURE 5-15 DEC mainframe VAX.

5.4 INTEGRATION COMPUTER CONTROL

The coordination of stations, transporter and storage computers is the primary responsibility of the integration computer, such as the VAX computer (Fig. 5-15). This coordination is achieved through the use of control algorithms for which related information is passed through a communications network. These are described in the following sections below, after an introduction to the FMS communications networks.

5.4.1 Communications Networks of FMS

In the early development of FMS, each control computer communicated only with the integration computer. Each CNC, PC or other type of controller had a single communication link with the central computer, forming a star pattern network (Fig. 5-16). In this topology, all communications passed through the central computer and no provisions were made for one CNC to communicate directly with another CNC or the transporter control. Some exceptions have been made which permit the CNC to have direct input and output with the

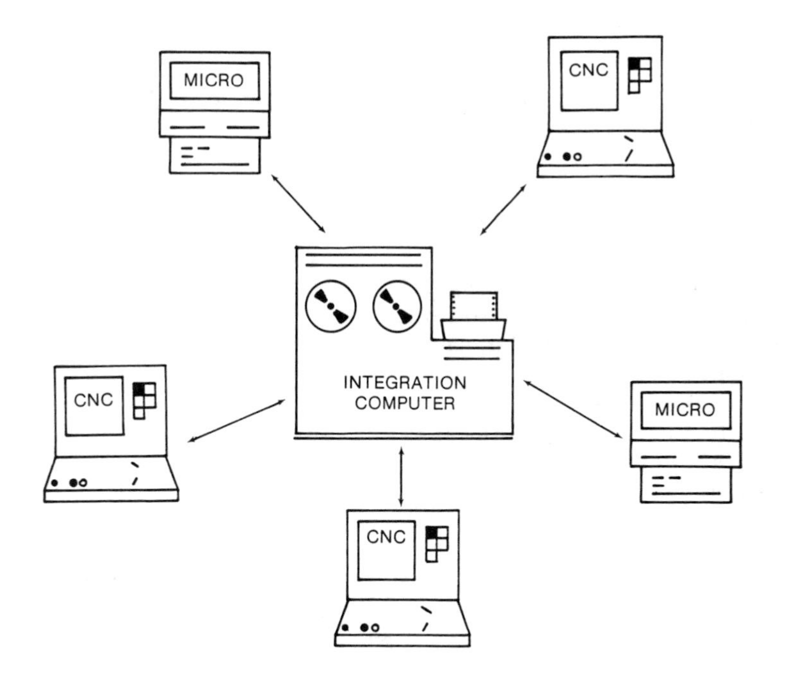

FIGURE 5-16 STAR topology.

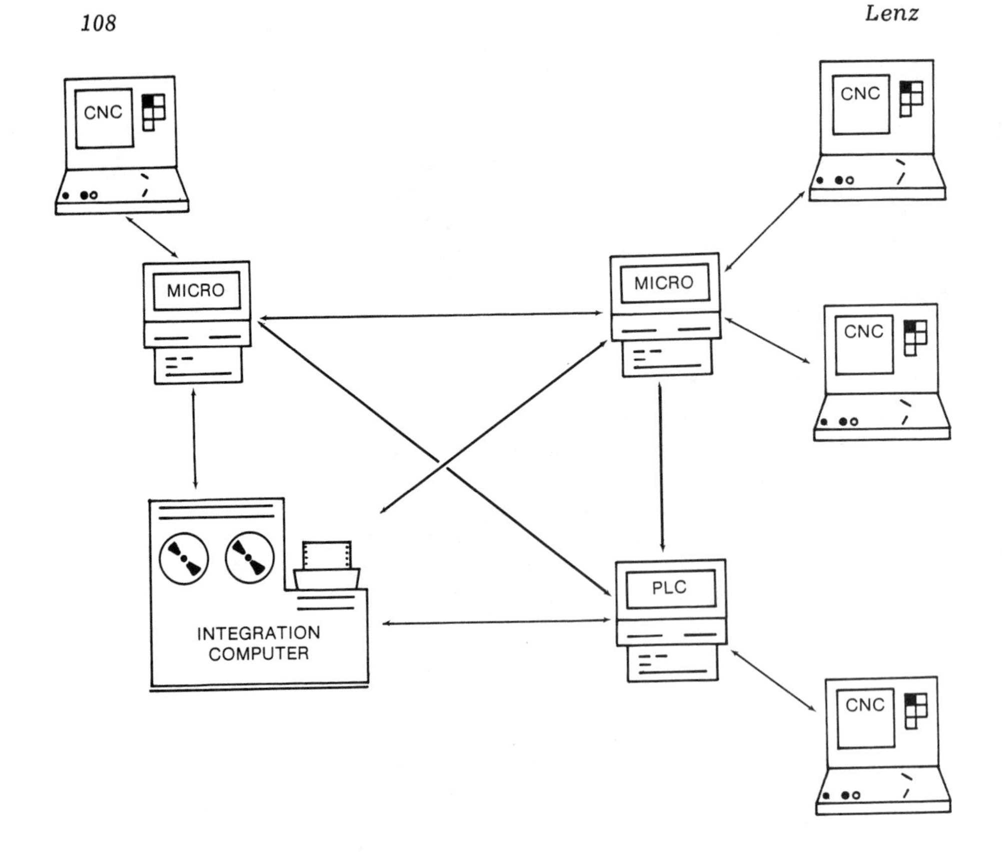

FIGURE 5-17 Network topology.

transport control. This avoids long delays between pallet transfers due to lengthy response time of the integration computer. However, such additions are unique to the controls and require specific compatibilities between the control computers.

Recent developments have led to network type communication structures. In the network pictured in Fig. 5-17, each control is a node along a single communication cable. For such a topology to work, however, each control on this cable must be able to talk the same protocol. That is, the method for talking to the transport PC must be identical to the method for talking to a CNC. Allen Bradley was one of the first to announce a communication network which allowed PCs and CNCs to communicate with each other.

These networks enable PCs and CNCs to communicate as long as they are from one vendor. However, it is not uncommon for a single vendor to deliver all controls for an FMS; consequently,

*** TOOL STATUS ***			
STATION # 7			
TOOL ID	POSITION		LIFE REMAINING %
1" DRILL	1	4013MIN	25
1" TAP	2	3627MIN	46
1 1/2" DRILL	3	5200MIN	16
1 1/2" DRILL	4	433MIN	78
1 1/2" TAP	5	1888MIN	67

FIGURE 5-18 Tool status report.

vendor supplied networks did not have many instances for application within the FMS. One development which promised to resolve this communication network problem was the introduction of Manufacturing Automation Protocol (MAP) by General Motors. Its intention was for all CNC, PC and computer manufacturers to provide communication capability with MAP. This type of network will have wide application in the FMS.

5.4.2 Management Information

Besides the coordination of all control computers, the integration computer provides a single point for management information reporting. The MIS reports fall into two categories: operational and performance.

Operational Reports

The operational reports provide timely data for the minute to minute operation of the FMS. These reports include fault detection with appropriate signals to operators, tool status reporting expected life remaining on each tool (Fig. 5-18), current pallet positions (Fig. 5-19) and current station activities (Fig. 5-20).

Because of the prompt need for this type of information, these reports are available through terminal request to the integration computer and are usually displayed directly on the terminal. One type of report which summarizes station status, pallet status and

*** PALLET STATUS ***

PALLET #	PART TYPE	SERIAL #	CURRENT STATION	CURRENT OPERATION	NEXT STATION	NEXT OPERATION
4	AB234	034	7	10	8	13
5	AB234	035	3	5	4	7
7	C1000	1349	2	3	3	4
10	F200	24	10	11	2	5

FIGURE 5-19 Pallet status report.

*** STATION STATUS ***

STATION	CURRENT PALLET	FAULT #	LAST START TIME
3	5	0	04-03-86:12:32p
4	1	0	04-03-86: 1:06p
5	2	0	04-03-86: 1:42p
6	1	53	04-03-86: 2:01p

FIGURE 5-20 Station status report.

transporter status is the dynamic line display. This display is a graphic display of the FMS layout on a computer terminal with the use of images to represent the vehicles, pallets and stations. Colors, position on the screen and text are updated to correspond with the status of the FMS hardware.

Performance Reports

The second area for MIS reporting is in FMS performance. Performance measures include parts required, completed and in-process; station utilization including major downtime reports; part inspection data if collected during the process; and transporter utilization. One report which summarizes much of the performance information is the *Part Trace Report*. This report (Fig. 5-21) contains a printed list of each part which has passed through the system. With each part is a list of all the operations which took place, the respective station number and the beginning and ending date and time. If the part has multiple sequences, this report should print information on all sequences with respective pallet numbers. This report provides a complete record of the part's travels and how long it spent at each station. It is also useful to have the elapsed time for each operation so it can be compared to the expected operation duration. This provides an easy method for identifying the occurrence of a major fault during part operations.

The part trace report provides a great deal of information on the system performance but is also used for quality assurance. When a part has been inspected and found to be out of tolerance, a reasonable explanation is needed. This requires tracing all action that the part took while in the FMS. The problem could be due to

*** PART TRACE REPORT ***

PART TYPE AB2

SERIAL # 00215A

OPERATION #	PALLET #	STATION #	START TIME	END TIME	ELAPSED TIME
1	3	3	11:12a	11:30a	19
2	3	4	11:40a	11:58a	19
3	3	6	12:10p	12:30p	21
4	3	10	12:46p	1:03p	18
100	10	2	1:15p	1:34p	20
110	10	4	1:46p	2:00p	15
120	10	10	2:10p	2:22p	13

FIGURE 5-21 Part trace report.

a number of reasons but the part trace report will contain all necessary information to provide a complete record of the part's movements throughout the system. This will include all pallet numbers used, a list of operations and the start and finish times for each. Elapsed time calculations are useful for detection of abnormal operation times, which can direct attention to the fault report at the specified time.

In order to produce a part trace report, a single source for all part activity is needed. This is one primary role of the integration computer. The other major function is to coordinate all activity in the FMS and to provide "optimal" efficiency in its operation. In the coordination process, integration computer programs must make intelligent decisions as to when and where parts are to be most beneficially transported. No one formula or priority calculation can be used to make such diverse decisions, so each decision should have its own algorithm.

The following is a description of seven decisions the integration computer must make with a list of alternative algorithms for each.

5.4.3 Part Balance Control

Part balance control determines which part is to be introduced into the FMS. This decision is made whenever a load station, pallet and raw part is available. The objective in making this decision is to schedule parts into the system to meet some planned production requirement. Proper control attempts to provide uniform station loading without overloading in-process storage facilities. In many cases, station loading is determined by the number and type of pallets which are active in the system, which may vary for different production schedules. Some examples of part schedule algorithms are given in Fig. 5-22. Part III will contain more details of FMS design.

Operation Sequence Control

Operation sequence control determines the sequence of operations for a part. The sequence is often restricted to a predetermined order because of fixturing, part geometry and part programming constraints. This decision is made after each operation is completed and is used to determine which station should be selected for this next operation. Some algorithms for this decision, shown in Fig. 5-23, include: fixed sequence, intermediate buffering or independent sequence based upon station availability. Fixed sequence implies that only one order exists in which the part can have its operations performed. Intermediate buffering uses the fixed sequence but allows the pallet to travel to designated storage areas to alleviate congestion when its next station is not available. The choice of where to buffer and for how long is determined by the integration

→ → Batch Scheduling

Parts are run with limited mixing of different part types.

→ → Production Requirements Mix

Parts are scheduled to meet all requirements on a uniform basis.

→ → Part Sequence

Parts having second and third fixtures (while passing through the system) can not be scheduled until previous sequences have been completed.

FIGURE 5-22 Part scheduling algorithms.

computer algorithms. A third algorithm allows for an independent operation sequence based upon station availability. This algorithm provides the most flexibility but overloading of some stations can be expected when parts require their last operations. Part III includes a discussion of design relating to operation sequence flexibility.

→ → Fixed Order

Alternative Order Based Upon Station Availability

→ → Buffering

Parts do not travel from station to station but can be stored temporarily while a station is unavailable.

FIGURE 5-23 Operation sequence algorithms.

→ → Idle Station

→ → Lowest Backlog

Backlog is the total minutes of work already assigned to the station.

→ → Closest Empty Queue Position

FIGURE 5-24 Station selection algorithms.

Station Selection Control

Station selection control determines which station is best suited to perform the next operation. This decision is made after a part has just completed an operation and is often merged into the decision of finding the next operation. Algorithms for this decision, illustrated in Fig. 5-24, include selection of the idle station, station with adequate tool life, station with the lowest backlog or station with in-process storage availability. The most common of these algorithms is the Backlog Selection Rule because it allows for idle stations having high priorities and incorporates storage availability to some degree.

In-Process Storage Control

In-process storage control determines when and where to store a part temporarily to prevent blocking at a station or an interruption in material flow. The flexibility of this control is subject to the amount of storage which is available at the station or at other locations accessible by the material handling system. This decision is made when a part has found its next operation and station and has requested transportation. As illustrated in Fig. 5-25, algorithms

→ → Due Date

→ → First In, First Out

→ → First In System, First Out

FIGURE 5-25 In-process storage control.

FIGURE 5-26 Transporter selection control.

for this decision include the part occupying its current position until the next station has availability, taking the part to dedicated storage facility, or keeping the part in the material handling system until the next station has availability. All of these algorithms attempt to minimize the congestion of material flow, but are limited by the capabilities and availabilities of the FMS hardware.

Transporter Selection Control

Transporter selection control determines which transporter is best suited for transporting a part from its current location to its desired location. This decision is made whenever a transporter is available and a part is requesting transportation. Algorithms for this decision include assignment of the idle cart, closest cart or feasible cart (Fig. 5-26). The feasible cart algorithm is used when vehicles are restricted in the area in which they can serve.

Traffic Control

Traffic control determines the traffic analysis to prevent collisions and deadlocks in the material handling system. Because of response times and the amount of communications required, these decisions are best made by the transport control computer and not the integration computer. However, traffic control requires a review of all parts and their destinations, which is information the transporter computer might not have. Algorithms for this decision, such as those in Fig. 5-27, depend upon the FMS layout and include bidirectional control, push—pull control, trunk line control, two-pallet position vehicle control and loop with spurs control. The push—pull control is used for many vehicles on a unidirectional path. This control uses zones and allows no more than a single vehicle in the zone. Figure 5-28 illustrates the trunk line control used for an AGV where all stations are located on spurs which connect to a bidirectional trunk line. In this control, use of vehicles is anticipated and they are positioned by the station before they are actually needed. This pre-positioning is more commonly referred to as "look ahead" because the decision is based upon anticipation of future events. The accuracy of this decision is tied to the accuracy of

→ → Bidirectional Vehicles

No restriction of service area.

→ → One-Way Loop Layout

All one way track in any topology.

→ → Trunk Line Layout

All one way track with each station located on a spur from the trunk line.

→ → Two-Pallet Position Vehicle

→ → One-Way Loop With Bidirectional Spurs

Vehicle can leave main track to service a station.

FIGURE 5-27 Traffic control algorithms.

predicted future events. The two pallet position vehicle control (Fig. 5-29) attempts to swap parts between vehicle and station. This control is useful when parts want to exchange positions such as a system with load/unload and identical machining centers. In this case, parts at the load/unload require a machining center next, and parts at the machining center require a load/unload next. When

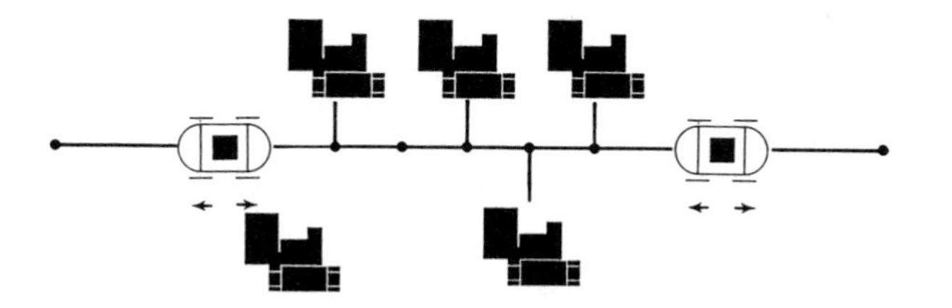

FIGURE 5-28 Trunk line vehicle system.

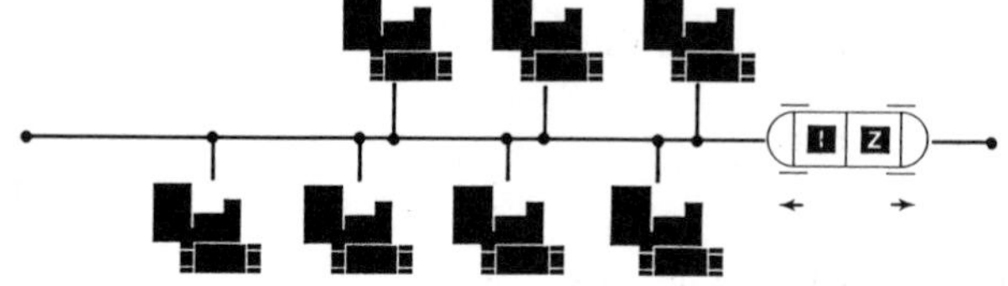

FIGURE 5-29 Two pallet vehicle system.

this condition arises, optimal use of the two pallet position vehicle can be made. Figures 5-30 through 5-32 show other examples of traffic control algorithms, each dealing with specific types of material handling hardware.

Tool Management Control

Tool management control determines when tools are to be replaced in each of the stations. This replacement can be due to changes in the production or to individual tools wearing out from use. In the case of production changeover, the mix of parts active within the FMS has changed, and the tool capacity in the stations must be adjusted for this change in production requirements. With machining centers, this can be accomplished by changing the tool configurations in each of the stations. Algorithms for changeover of tools, such as those seen in Fig. 5-33, must use the current list of tools in each station and have the required list for each station for the revised production. When it has been determined that a station is no longer needed to perform an operation from a previous production schedule, the set of tools which need to be replaced can be done so manually or automatically by means of a tool pallet. To

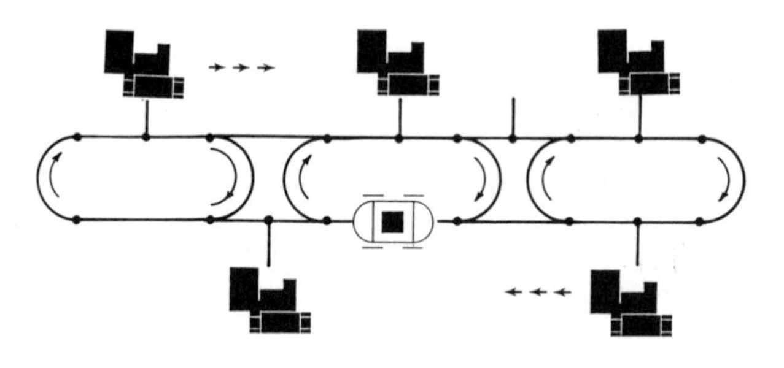

FIGURE 5-30 One-way loop vehicle system.

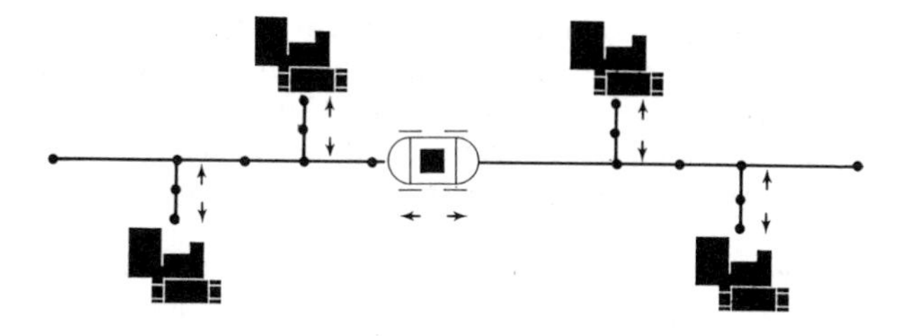

FIGURE 5-31 Trunk line with bidirectional spur.

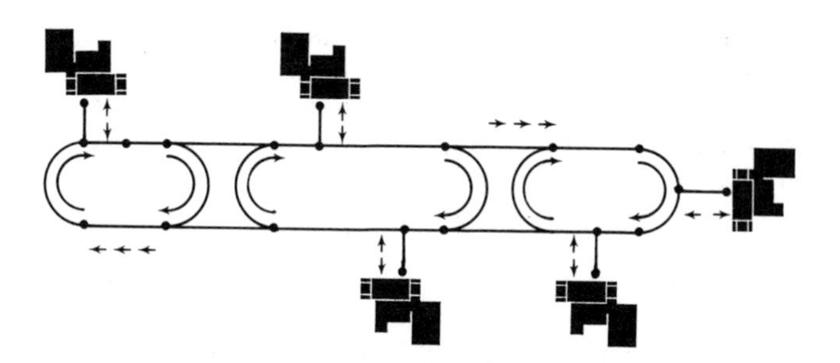

FIGURE 5-32 Loop vehicle bidirectional spur.

→ → Single Tool Replacement

When a tool reaches its warning point replace only that tool.

→ → Magazine Replacement

Complete replacement of all tools

→ → Periodic Replacement of Back-Up Tools

FIGURE 5-33 Tool management algorithms.

ensure that parts do not arrive before the tools have been delivered, the station should not be added to the route file until it has adequate tooling for the necessary operations. In this way, the part route files which contain the list of feasible stations must be adjusted automatically by the integration computer as it records status of tool changeover.

The second reason for tool replacement is when a tool has surpassed its wear life. Some algorithms for resolving this situation are to use a backup tool in the tool changer and wait until several tools need to be replaced before replacing any. Another method is to take the station off of part route lists and only let it perform operations for which it has adequate tooling. If neither of these two algorithms are used, then it is possible to continue to use the tool while sending warning messages to signal that an exhausted tool is being used. The reason for this final algorithm is that the tool data collection is usually static and may not be current with the actual situation. The data may be unreliable; therefore, it is advisable to use a tool until a second signal is given which indicates that the tool is really worn.

5.5 CONTROL HIERARCHY

The control for the FMS often provides the integration of its many "islands" of hardware. This control consists of a hierarchy of computers (Fig. 5-34), ranging from microprocessors, which control a single point for coordination. In this hierarchy, control decisions

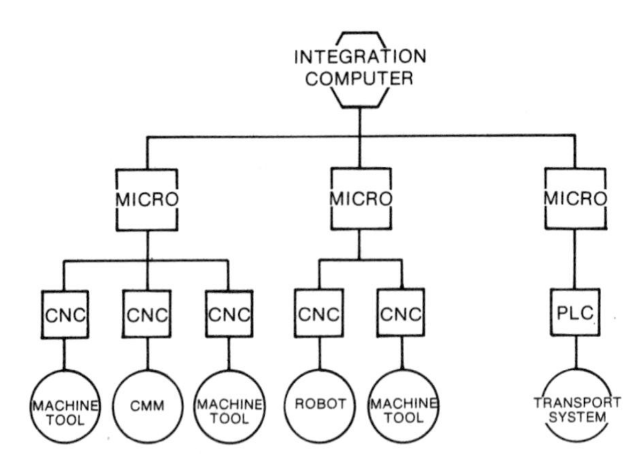

FIGURE 5-34 Control hierarchy.

are made at all levels so the central computer provides more coordination activity, rather than a single source for all decisions.

When this hierarchy is not present, usually due to the weakness of low level computers in handling data, communication and decisions, the central computer becomes the default where all decisions must eventually be made. The problem with this control design is that interruptions can be expected in motion to allow time for the integration computer to receive communications, make a decision and transmit the command through communication. The only means to avoid these delays is to offload this type of decision to computers which are directly connected to hardware. In order to accomplish this, the controls might need to establish direct communication with each other, rather than relying on a common link through the integration computer. This will require expanded capabilities beyond those normally needed to control the component as an independent island. The establishment of distributed decision making and of the communication network is essential for the effective coordination and integration of the many islands in FMS.

5.6 CONCLUSION

The control components in the FMS consist of a hierarchy of programmable controllers, CNCs, microcomputers and integration computers. Each has specific areas of control and carries out a subset of the decisions necessary to automatically operate the FMS hardware. The coordination of these decisions is accomplished by common values of component status across a network. With the use of a network, distributed decision making and hierarchical control, the control for an FMS requires the state-of-the-art in computer technology.

When these components are understood, the next logical step is in design procedures and guidelines for the FMS. These are presented in detail as a three-step approach involving capacity planning, integration and operation. As stated in Chapter 1, automation requires that adaptation of a system to changes be done at the design/configuration level. Therefore, a thorough understanding of FMS design is essential to understanding the capabilities of the new technology.

6

Aggregate Planning for Flexible Manufacturing Design

Aggregate planning of a flexible manufacturing separates the capacity problem of design from the integration problem. This separation allows utilization to be computed for stations which become the integrated system's "targets." Without separation, performance evaluation must deal with capacity and integration effects simultaneously. The role of capacity planning is to provide a quick, easy way to estimate system performance from forecasted production and process data.

Aggregate planning of flexible manufacturing requires use of a mathematical model focusing on the integration of components. One of the first such models was CAN-Q, the Computer Analysis of Network Queues. This model utilizes a queuing theory model with adaptation for the integration of many such queues. It provides an evaluation of system throughput without a large computer or extensive data. Even though the weakness of such a tool is its necessary assumptions, the results can be quite accurate for predicting system capacity.

Since CAN-Q, many other design tools have become available from university research and commercial product development. One such program is the System Planning of Aggregate Requirements (SPAR) program which includes an automatic diagnostic and adjustment of unfeasible system designs. This program provides the capacity analysis of flexible manufacturing and is a data preprocessor for the MAST Simulation program. Both SPAR and MAST are available from CMS Research, Inc. in Oshkosh, Wisconsin.

No matter which planning tool is used for the study of flexible manufacturing, they all have similar roles in the design problem. This role or objective is to identify sufficient system capacity to meet a given production requirement. In establishing system capacity, such decisions as to the number of stations of each type, the number of vehicles and number of pallets must be computed from the requirements. Before the actual process of aggregate planning is described, a short discussion of relating these numbers to reality follows.

How can estimated data be used to reflect or predict actual flexible manufacturing performance? In other words, how can numbers inside of a computer possibly reflect a real systems operation? This question is important to answer before exploring the mechanics of aggregate planning. One must first identify those items which are common to both the computer numbers and the real system. The common characteristics, known as system performance measures, are station utilization, transporter utilization and flow times. The only way to tie the mathematical world to the real world is to ensure that realistic measures are used in both cases. Station utilization, which can realistically be achieved during the planning horizon, should not be exceeded in mathematical calculations. To permit a mathematical result to show 100% station utilization will not result in useful operation information because the 100% prediction is not realistic. This also applies to transporter utilization and flow time. It is not realistic to have flow time equal only to total operation time because any pact will be delayed for some time while waiting for a busy station.

When a planning tool is used to compute utilizations or flow times, the values which are received must be realistic targets for the actual system. If they are not, then there is no relation between the numbers inside of a computer and a real system. When the designer incorporates realistic design targets into the mathematical evaluation, system capacity can be derived from mathematical results. The role of computer simulation is to then study the effects of integration on each component.

No matter which aggregate planning tool is used, the data requirements are similar. This data includes a list of parts, production requirements, a process plan and some planning horizon. From this input, the number of stations, transporters and inventory level can be determined. A discussion of the data input needed for aggregate planning of flexible manufacturing follows.

6.1 PLANNING HORIZON

The planning horizon represents the amount of time allotted for the completion of the required production. This time is computed from the hours per shift and the number of shifts which are available to complete the required production. In some cases, further detail may be given such as the number of shifts per week and the number of weeks in the planning period. However, because individual companies may have different definitions of what constitutes a month or a year, these units are not usually used. Once the planning horizon has been established, it is assigned as the capacity of each component in the system. For example, if a planning horizon of 80 hours was calculated, it is assumed that each station or transporter in flexible manufacturing is available for 80 hours during the planning period.

Almost anyone with knowledge of machinery in operation will argue that no component could be available for the full production period. Because of the possibility of breakdowns, lost time due to changes in shift, or shutdown and start-up, it is necessary to reduce the component's available time. This is done through the use of efficiency factors.

An efficiency factor is the maximum percentage of time any component can expect to be available during the planning period. Efficiency factors will vary depending upon the degree of integration between the system components. For example, because components within the transfer line are completely integrated, an efficiency as low as 70% might be used, whereas in flexible manufacturing with more independent operations between components, an efficiency of 80% to 90% might be used. These efficiency levels result from experience with the actual running of similar equipment and time required to alleviate accompanying problems. To apply efficiency factors in the planning of flexible manufacturing, the capacity or time available for each component is adjusted to reflect the efficiency level. As Fig. 6-1 illustrates, in the case of a planning horizon of 6000 hours with an efficiency factor of 80%, the planning horizon would be reduced to 6000 × 0.8 or 4800 hours. Even though this is the traditional method for establishing component capacity, it does have some side effects on the design.

One side effect is that computed utilizations will be higher than those with the longer planning horizon. This is not necessarily bad as long as the efficiency factor is taken into consideration and used to adjust to an overall utilization, such as the example in Fig. 6-2. However, by reducing the planning horizon, pressure is placed upon the system to complete all required production in some time period less than the total planning period. This raises the expected production rate to unrealistic levels, which correspond to aggregate planning results of 100% station utilization.

One hundred percent station utilization alone is not the problem. It only results because of mathematics and is really no larger than the selected efficiency factor. Where it does become a problem is in establishing the amount of work in-process for the system (i.e., number of parts). The design tool attempts to provide enough

8 HOURS/SHIFT	
2 SHIFTS/DAY	
5 DAYS/WEEK	
50 WEEKS/YEAR	
50 WEEKS/YEAR = 6000 HOURS	
EFFICIENCY LEVEL	TOTAL HOURS
100%	6000 HOURS
80%	4800 HOURS
60%	3600 HOURS

FIGURE 6-1 Planning horizon vs. efficiency factor.

work in-process to maintain target levels of station utilization. If this is 100%, as shown in Fig. 6-3, the number of parts required to maintain this level will be suggested for the system. Even though the stations are not overcapacitated, the amount of work in-process will reflect a level which is necessary to "consume" 100% of the station capacity.

STATION A

TOTAL REQUIREMENT IS 4600 HOURS

EFFICIENCY LEVEL	TOTAL HOURS	% USAGE
100%	4600 ÷ 6000	76%
80%	4600 ÷ 4800	95%

FIGURE 6-2 Efficiency factor vs. station utilization.

STATION UTILIZATION	# OF PARTS NEEDED IN SYSTEM	FLOW TIME
100%	10	76 MINUTES
80%	7	55 MINUTES
60%	4	40 MINUTES

FIGURE 6-3 Station utilization vs. work in-process.

This planning for 100% station utilization has been the most common problem of flexible manufacturing design. It has resulted from the violation of a fundamental rule of manufacturing: the amount of work in-process must be balanced with the capacity of the machinery. If a machine is 100% available but only scheduled for 80% usage, it will appear to be underutilized because of a lack of work. When the amount of work is scheduled for 100% but the machine is down 20% of the time, a queue of work accumulates for this machine. In a job shop or nonintegrated layout, this surge of work can either collect on the shop floor or be taken to some designed storage and retrieval area, but it will not have any carry-over effects. Within flexible manufacturing, work in-process must occupy some space in the system, and when an imbalance arises between work level and station capacity, integration effects will occur. These will lower the actual station use to below its level of availability.

There is nothing wrong with planning for 100% station utilization, but this amount of work in-process can only be required when the system has full level capacity. As individual components fail and are repaired, the system capacity will vary. Because the system capacity varies, the level of work in-process must be adjusted so balance remains at all times. This adjustment is defined with the WIPAC Curve, which was described in Chapter 3.

In the past, systems have been designed to 100% station utilization with a corresponding level of work in-process. But when the system is installed, it operates at some level less than this (usually near efficiency level) and poor system performance is observed. When the system does not perform up to the design level, it is usually because it was never required to do so. But the amount of work in-process causes integration effects, preventing the station use from reaching its level of availability. The establishment of work in-process levels for various station capacities is an important part of flexible manufacturing design. Its computation is described later in this chapter.

6.2 PRODUCTION REQUIREMENTS AND PROCESS PLANS

Once the planning horizon has been established, the next area for data input is in the part descriptions. These descriptions include the production requirements, a process plan, and some factor for in-process storage estimation. The production requirement in Fig. 6-4 is the number of pieces for each part which is to be produced within the planning horizon. This number might need to be adjusted to include scrap factors, reworked percentages, spare parts and the number of parts which are simultaneously processed.

In aggregate planning, the assumption is to compute requirements based upon a uniform need of the production during the planning period. This does not account for the effect of batching parts and the need to produce parts at a higher rate than a uniform rate over the entire planning period. Batching of parts is considered an operational type problem and is best studied with a computer simulation where aggregate planning is used to establish overall system capacity.

The second data input in the part description is the process plan (Fig. 6-5). The process plan consists of a sequence of operations along with some estimated duration for each. The operation durations must account for the number of parts per pallet and reflect the equivalent part program time. Depending on planning programs, pallet exchange times will be added to operation durations or contained as part of the station description.

The third data input for each part is the amount of time a part could expect to spend in storage waiting for an operation or for transportation. In CAN-Q, this value is not needed as input because the model estimates queue lengths and can compute queue durations. However, in other planning tools, an estimate for this

PART #	TOTAL REQUIRED
A1324	500/YEAR
C17C34	1800/YEAR
1753	2400/YEAR
44D31	2500/YEAR

FIGURE 6-4 Production requirement.

PART # 1753

PROCESS	TIME
LOAD PALLET	3.12 MINUTES
ROUGH MILL	10.17 MINUTES
SEMI FINISH-DRILL	15.37 MINUTES
DRILL, TAP	8.3 MINUTES
FINISH BORE	14.3 MINUTES
UNLOAD PALLET	2.0 MINUTES

FIGURE 6-5 Process plan.

time is needed to determine a flow time duration. (Section 6.8 contains some guidelines for estimating this value.) Figure 6-6 shows an example of computing a parts flow time.

Aggregate planning programs require a variety of additional data, ranging from shuttle times for each station to the average

PART #1753

TOTAL PROCESS TIME	53.26 MINUTES
TOTAL TRANSPORT TIME (5 MOVES @ 2 MINUTES EACH)	10 MINUTES
TOTAL STORAGE TIME (1/2 OF PROCESS TIME)	26.63 MINUTES
PART FLOW TIME	89.89 MINUTES

FIGURE 6-6 Part flow time.

STATION TYPE	NUMBER AVAILABLE	TOTAL MINUTES		UTILIZATION
		AVAILABLE	REQUIRED	
LOAD	1	2880.0	805.0	28.0
MILL	2	5760.0	4280.0	74.3
DRILL	1	2880.0	2710.0	94.1
UNLOAD	1	2880.0	805.0	28.00
BORE	2	5760.0	3675.0	63.8

FIGURE 6-7 Station utilization.

time required for a part to be transported between successive operations. With this data, the aggregate planning tools are used to compute requirements in each of three areas and to compare them to capacity levels. These areas of system evaluation are station utilization, transporter utilization and inventory requirements.

6.3 STATION CAPACITY AND UTILIZATION

The primary output of aggregate planning for flexible manufacturing is to identify the targeted station utilization (Fig. 6-7). The utilization must include all the time required to perform operations on all necessary production and also must account for part changing between operations. This computed number represents the minimum station utilization level necessary to complete the required production in the planning horizon.

When the planning horizon has been adjusted to account for component efficiency, the station utilization will be higher thanwhen the entire planning horizon is used. Often this higher value will exceed realistic levels. The alternative method is to leave the planning horizon at its full value so that each station has capacity for the entire production period. When using this method, the station utilization value cannot exceed the efficiency factor. This second method is more closely related to how the real system will eventually operate. Each station will be part of the flexible manufacturing facility for the total production period, but due to interruptions its usage will approximate some efficiency level.

Using the total planning period and keeping station utilization below the efficiency level not only reflects actual system operation but also establishes a corresponding level of work in-process which will be much more realistic than the level to maintain 100% station utilization.

6.4 TRANSPORTER CAPACITY AND UTILIZATION

As in the case of station capacity, it is assumed that each transporter has capacity equal to the planning horizon. The requirement of the transporter is computed by accumulating the number of transport moves times the average cycle time for a vehicle over all production. This utilization value, pictured in Fig. 6-8, is estimated because the average cycle for a transporter is not known with certainty. Here is a good application for computer simulation. In the simulation, each transporter can be simulated in its detailed motion and the average cycle can be reported as part of the summary statistics. But for aggregate planning, this value is estimated to

NUMBER TRANSPORT	TOTAL MINUTES AVAILABLE	TOTAL MINUTES REQUIRED	TRANSPORT UTILIZATION
1	2880.0	6780.0	NOT FEASIBLE
2	5760.0	6780.0	NOT FEASIBLE
3	8640.0	6780.0	78.4

FIGURE 6-8 Transporter utilization.

give close approximation to the number of vehicles and expected level of activity for each.

6.5 STORAGE CAPACITY AND INVENTORY REQUIREMENTS

The third area for flexible manufacturing capacity planning, charted in Fig. 6-9, is in the work in-process level necessary to maintain targeted station and transporter utilizations. These results are computed from the estimated storage time a part will spend while active in the flexible manufacturing facility. As was described,

PART ID	FLOW TIME	% STATION	% STORAGE & TRANSPORT	CYCLES PER HORIZON
A1324	231.0	45.5	54.5	12.5
C17C34	338.0	45.9	54.1	8.5
1753	181.0	44.2	55.8	15.9

FIGURE 6-9 Inventory requirements.

some guidelines are available for "optimal" design but these results are based upon estimated flow times. The equation of station time plus transport time plus storage time yields the flow time, previously seen in Fig. 6-6. Because estimated values are used, the results become approximations for which computer simulation can give accurate measures.

Once the flow time has been estimated, the number of cycles which are available within the planning horizon can be found, and eventually the inventory level part needed to produce the required production can be computed. In many aggregate planning tools, the production is assumed as uniformly produced throughout the production period and an average number of parts will be used. When batching is required, the system has to be able to produce specific parts at a greater rate and thus, higher levels of work in-process will be needed. This greater percentage of one part will replace other production requirements in the short term, but addition or replacement of parts depends upon batch size and variability in process plans between the parts. This type of problem is best solved using computer simulation.

6.6 TOOL CONFIGURATION FOR FMS MANUFACTURING

Some aggregate planning tools for flexible manufacturing include the ability to collect data regarding each operation and the list of individual tools that are used in that operation. This information can then be used to establish a feasible tool list for each station and assist in deciding which parts can run simultaneously in the flexible manufacturing facility.

Once the individual tool time, number of positions required in the tool changer and an estimated life for each tool in an operation have been collected, several aggregate reports regarding tool configuration are available. These include a tool list for each station, a tool required report and a tool configuration containing sister tools for all stations.

6.6.1 Tool List

Figure 6-10 depicts the tool list report which contains a list of unique tools needed for a station based upon the process plans of the required production. This report can be used to show what common tooling exists between various parts and which parts are best suited for simultaneous production.

PLANNING HORIZON: 480

DRILL STATION

TOOL ID	PART ID	TIME
1" DRILL	PART 1753	5.20
	PART A1324	3.40
	PART 123	5.60
1" TAP	PART 1753	3.30
	PART A1324	4.20
	PART 123	3.40
1 1/4" DRILL	PART 1753	6.40
	PART 123	2.10
1 1/4" TAP	PART 1753	5.40
	PART 123	3.20

FIGURE 6-10 Tool list report.

6.6.2 Tool Required Report

The tool required report (Fig. 6-11) contains the total requirement of tools to produce all parts during the planning horizon. Utilizing the estimated tool life, this report can compute the total requirement for each tool and also find the number of tools. Because the tool life is an estimated value and depends upon usage conditions, the report will provide a target to the level of necessary tool activity. As actual data becomes available, the tool life estimates can be adjusted and this report will yield more accurate results.

6.6.3 Tool Configuration Report

The third major report from tool evaluation is the tool configuration report for each station (Fig. 6-12). In order to produce this report, the total tool requirement is used along with the estimated life and the number of tool positions. Before an actual configuration can be established, one additional item of data is needed. This is the length of time that a station will operate without an interruption

PLANNING HORIZON: 480

DRILL STATION
NUMBER OF STATIONS: 2 TOTAL CAPACITY: 960

TOOL ID	PART ID	TIME	POTS	TOTAL REQUIRED	PERCENT UTILIZATION
1" DRILL	PART 1	5.40	1	64.80	6.75
	PART 10	3.40		136.00	14.17
	PART 123	5.60		235.20	24.50
				436.00	45.42
1" TAP	PART 1	3.40	1	40.80	4.25
	PART 10	4.50		180.00	18.75
	PART 123	3.40		142.80	14.88
				363.60	37.88
1 1/4" DRILL	PART 1	7.60	1	91.20	9.50
	PART 10	2.10		84.00	8.75
				175.20	18.25
1 1/4" TAP	PART 1	5.50	1	66.00	6.88
	PART 10	3.20		128.00	13.33
				194.00	20.21
DRILL STATION RESULTS:			4	1168.80	121.75

FIGURE 6-11 Tool required report.

PLANNING HORIZON: 480 TOOL WINDOW: 480

TOOL ID	OPERATION ID	TOOL LIFE	TOTAL REQUIREMENT	TOTAL TOOLS	TOOLS PER STATION	POTS
1" DRILL	DRILL	50	436.00	8.72	9	9
1 1/4" DRILL	DRILL	35	194.00	5.54	6	6
6" CUTTER	MILL	100	256.80	2.57	3	9
8" CUTTER	MILL	120	371.20	3.09	4	12
10" CUTTER	MILL	20	370.40	18.52	19	57
1 1/4" DRILL	DRILL	50	175.20	3.50	4	4
1" TAP	DRILL	35	363.60	10.39	11	11

FIGURE 6-12 Tool configuration report.

PLANNING HORIZON: 480 TOOL WINDOW: 480

DRILL STATION
NUMBER OF STATIONS: 2

TOOL ID	TOOLS PER STATION	TOOL POSITIONS PER STATION
1" DRILL	9	9
1" TAP	11	11
1 1/4" DRILL	4	4
1 1/4" TAP	6	6
	30	30

PLANNING HORIZON: 480 TOOL WINDOW: 240

DRILL STATION
NUMBER OF STATIONS: 2

TOOL ID	TOOLS PER STATION	TOOL POSITIONS PER STATION
1" DRILL	5	5
1" TAP	6	6
1 1/4" DRILL	2	2
1 1/4" TAP	3	3
	16	16

FIGURE 6-13 Tool configuration summary with different windows.

due to tool shortages. This time interval, for which no tools need replacement at the station, is referred to as the tool window. Figure 6-13 shows the use of different tool windows in summary of the tool configuration report pictured in the previous figure. In many manufacturing systems, the tool window might be one shift. In this instance, once per shift, worn tools are replaced, ensuring that enough tool capacity exists to operate the system for one shift.

Once a tool window has been provided, a suggested tool list can be computed for each station. This list contains the number of tools which are needed to maintain uninterrupted service, and the total positions in the tool changer needed for these tools. This information yields the number of backup tools and positions which must reside in the tool changer. This report can then be used to establish the maximum tool window for a given tool storage capacity.

This tool configuration study assumes that some real time tracking of tool usage is performed as part of the control system. When a tool is observed to reach its life, a backup tool can be automatically activated to replace it. If the system lacks the capability for switching to backup tooling automatically, then manual intervention, which might as well include replacement of the worn tool, is required.

6.7 DESIGN GUIDELINES FOR AGGREGATE PLANNING IN FMS

Station, transporter and flow times can be computed from production requirements with the use of an analytic model for flexible manufacturing. Before these results can be useful in the flexible manufacturing design process, they must have some connection with performance of the real system. For this reason, use of analytic tools has given way to computer simulation as the reliable source for design data. But this need not be the case if proper guidelines are followed.

These guidelines can be categorized into three areas: station utilization, transporter utilization and congestion allowance. Each is summarized in Fig. 6-14 and described below.

6.7.1 Maximum Station Utilization

As previously stated, the only connection between reality and the analytic tool is component utilization. Proper use of an analytic tool requires that computed utilization for a component does not exceed a realistic level. Computed numbers must directly compare to a realistic expectation of station availability. Using an efficiency factor to reduce the planning horizon increases utilization beyond

STATION DESIGN

DESIGN CRITERIA:

Station utilization (part program execution) does not exceed 80% of available time.

RATIONALE:

Capacity for recovery from down components.
Uniform production levels.
Reduces impact of surging.
Maintain control of operation.

TRANSPORTER DESIGN

DESIGN CRITERIA:

Transportation utilization not to exceed 60% and average assignment time less than the shortest operation.

RATIONALE:

Parts wait for work tables not transporters.
Recoverability from periods of unreliability.
Transporters are not used for in-process storage.
Maintain control of operation.

IN-PROCESS STORAGE DESIGN

DESIGN CRITERIA:

Parts spend at least 60% of the time in the system at a station in operation.

RATIONALE:

Parts are queued for stations.
Balance operation times among stations.
Work tables are not congested because of congested queues.
Maintain control of operation.

FIGURE 6-14 Design guideline summary.

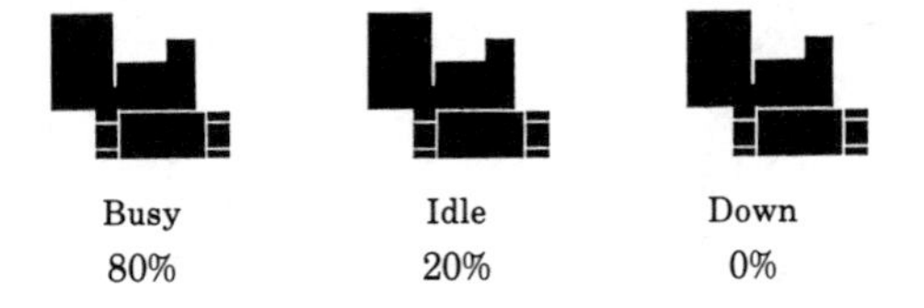

FIGURE 6-15 Target use areas.

reasonable levels and can be avoided by using planning horizons equal to the time available for facilities operation.

According to data taken from actual FMS performance, most stations operate between the 60% and 70% utilization level. Although this value is less than most designers desire, low performance is attributed to unavailability of stations and the integration effects due to inappropriate levels of work in-process and limited flexibility. In the properly designed flexible manufacturing where integration effects are kept to a minimum, actual station utilization can be close to station availability. As a result, production rates will be closer to planned (or expected) levels.

To further explain the role of integration effects upon planned station utilization, the difference between planning for unavailability and actual availability is explained. Planning with mathematical models will use an efficiency factor and provide the following target use areas, as seen in Fig. 6-15.

However, when the station becomes unavailable the percentage of down time will increase. What is desired is the trading of planned idle time for down time with no effect upon busy time. This will be typical in job shop or transfer line manufacturing strategies. But when integration effects exist, busy time will be affected by material shortages resulting from unavailable stations. In fact, the distribution of busy and idle time to down time may resemble the depiction in Fig. 6-16. From this, the station's availability is not below planned levels or consistent with efficiency factors. However, actual station use or production rate will be lower than targets due to integration effects. From this example, integration effects will actually cost 10% of production because it is not possible to simply convert all planned idle time to down time. The measurement and anticipation of these effects within flexible manufacturing are the subjects of the next chapter.

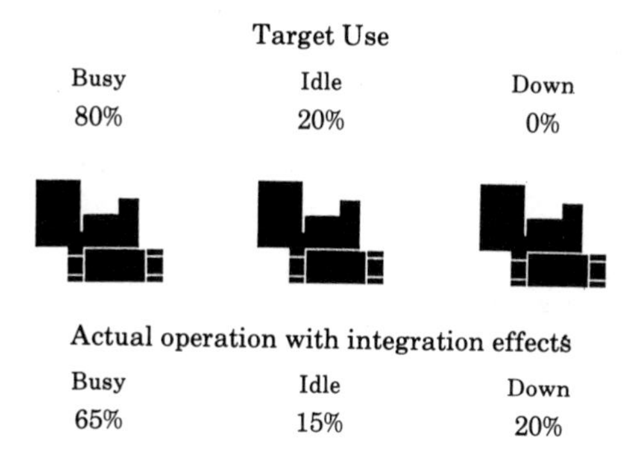

FIGURE 6-16 Target use areas with integration effects.

6.7.2 Maximum Transporter Utilization

A realistic target for transporter utilization must allow for technological improvements over the life of the facility. The transporter can be sized for the given production characteristics and its utilization can be as high as 75% without concern. However, as improvements are made in areas such as tool speeds and feeds, higher horsepower and faster mechanisms in the machine tool, operation times will decrease. As these cycles decrease, the frequency of part deliveries and pickups will increase at each station and this will result in an increased demand upon the transportation system.

Through the years of operation, the transporter requirement can be expected to increase. But unlike planning for station capacity, increases in transporter capacity must be part of the initial design. Because increases in transporter capacity are not directly proportional to the number of vehicles in the system, the addition of another vehicle does not ensure an equivalent increase in capacity. For example, an AGV system with five vehicles might be sufficient for current needs. But as cycles decrease or station utilization exceeds 90%, a sixth vehicle might be necessary. Adding another vehicle to the system will increase the traffic congestion, consequently increasing the average assignment time of the vehicles. This increased assignment time will offset the additional capacity of adding a vehicle. Therefore, the change in transporter capacity is determined by the increase in traffic congestion which results from the additional vehicle. In the extreme case, it might be possible that a five vehicle AGV might have greater capacity than a six vehicle AGV system.

Because additions to a transport system do not always increase its capacity by an equivalent amount, it is necessary to allow for capacity expansion during the initial design. One method is to design a transport system which has utilization not exceeding 60 percent. With this guideline, greater transporter capacity will be purchased than what is immediately needed. But this marginal cost will be much lower than the cost of increasing transporter capacity by redesign or changes to the system layout. In this case, an AGV is easier to install and is much more adaptable for expansion.

6.7.3 Inventory Requirements

Hos much time should a part be allowed to spend in flexible manufacturing? This is the most significant question because most of the benefits of flexible manufacturing are related to work flow control. Chapter 9 contains the discussion of flexible manufacturing benefits. Before a design guideline can be presented, however, a background in inventory requirements is necessary.

The time elapsed from the loading and unloading of a part is referred to as the flow time. This includes all operation, transport and storage time. Because of the material tracking system often found in the flexible manufacturing, the amount of storage time can be controlled just as operations and transportation are controlled. This ability to control the storage time is one of the primary benefits of flexible manufacturing. For example, in order to achieve a storage time of zero, a part would need to travel from its completed operation, and upon arrival move immediately into its next station to begin operation. The only way of accomplishing this is by keeping the next station idle, resulting in underutilization. In this situation, maximum production capacity could not be achieved.

The other extreme is to try to achieve 100% station utilization. In order to achieve this, a part would have to be ready to go into the station whenever it completes an operation. In this situation, parts might wait at a station longer than the required operation time. When this occurs, the part might spend more time in storage than in operation. However, excessive storage durations can also lead to high degrees of integration effects, which will create unpredictable performances in the flexible manufacturing.

The balancing of work in-process levels with station utilization levels is determined from the WIPAC Curve. This curve has an upward sloping portion where increases in work in-process levels are consuming greater levels of station availability. However, there will be diminishing returns, and greater increases in work in-process will be needed to obtain less increases in station utilization. This is easily observed as the curvature in the WIPAC Curve.

A short review of this curve follows, as it is important in establishing (quantifying) the relationship between flow time, work in-process level (inventory) and production rates (station utilization).

6.8 THE WORK IN-PROCESS AGAINST CAPACITY (WIPAC) GRAPH

Once the target station utilization, transporter utilization and inventory level has been computed, it is useful to quantify the relationship between work in-process, station utilization and flow time. This relationship was referred to as the triangle of integration in Chapter 3. An important part of any flexible manufacturing design is to quantify this relationship. The remainder of this section contains a procedure of estimating the WIPAC Curve using mathematical models. Chapter 7 contains the procedure for establishing the actual curve.

The WIPAC Curve is the locus of points representing the minimum work in-process level which can utilize the system at its level of available capacity. At this level of work in-process, the capacity of the system will be fully utilized and congestion will be at a minimum. Part IV contains the reasons why this characteristic lends itself to maximum benefits from integration. As the capacity of the system increases, more work in-process is needed to utilize the capacity. However, as the level of in-process parts increase, so will their flow times. As described in Chapter 3, the increase in the number of in-process parts will have a positive effect upon station use but the increased flow time will have a negative or counteractive effect. The WIPAC Curve is the tool by which these offsetting effects can be quantified and represented.

A sample WIPAC Curve is shown in Fig. 6-17. The work in-process level is represented on the horizontal axis and system capacity is represented along the vertical axis. System capacity can be expressed in terms of station utilization or parts per hour because these are equivalent except for their units of measure. In some evaluations, it might be easier to obtain data for parts per hour than to obtain data for station utilization.

Because the WIPAC Curve is a means to quantify the relationship between work in-process, station utilization and flow time, the question arises as to why the WIPAC Curve is only two-dimensional. The answer is that the work in-process level and flow time are functionally related by use of the storage factor, as depicted in Fig. 6-18. Because the flow time and work in-process are functionally dependent, only one needs to be graphed against the system capacity.

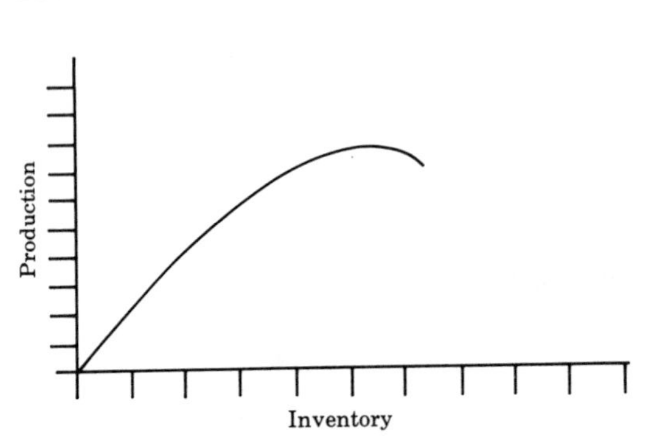

FIGURE 6-17 Sample WIPAC Curve.

The actual WIPAC Curve will be nonlinear. This characteristic results because of the counteractive effect which inventory and flow time have upon system usage. As inventory increases, so will the system use. Where the flow time is constant, the WIPAC Curve will be a straight line (Fig. 6-19). At some point, flow time will increase, which will have some counteractive effect upon system usage. This offsetting will cause the curve to bend downward. The increase in work in-process level will have its positive effect on station use reduced by the effect of increased flow time.

In the use of mathematical models, the relationship between work in-process and system use can be measured. With use of a selected storage factor, a flow time can be determined. But there is no precise mathematical method to determine how much the flow

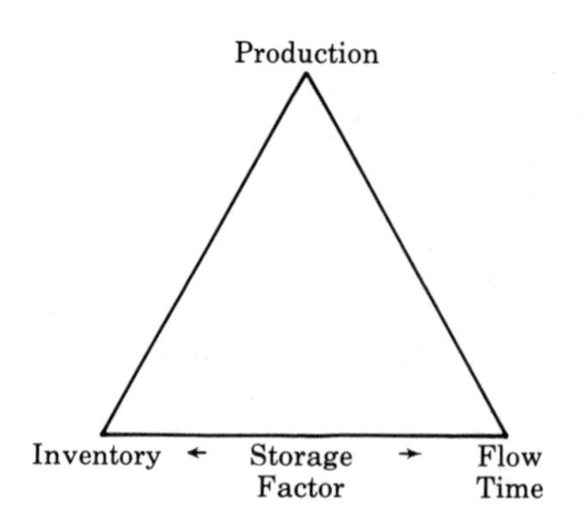

FIGURE 6-18 Storage factor definition.

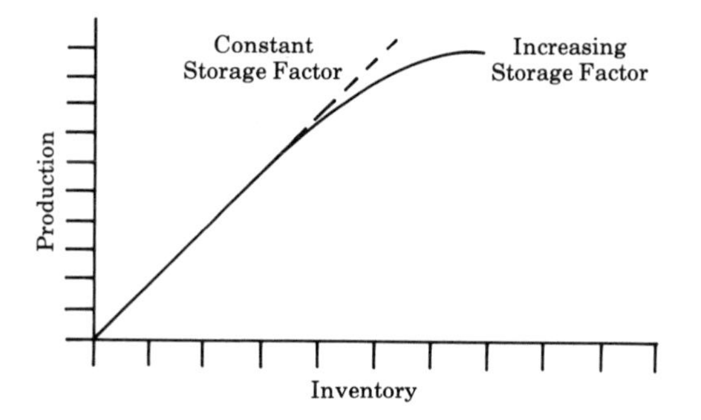

FIGURE 6-19 WIPAC graph with constant and variable flow times.

time will increase (i.e., how the storage factor will change) as the work in-process is increased. However, there are heuristic methods, such as CAN-Q, which can approximate the flow time for given inventory levels. The exact relationship between flow time and work in-process levels can only be determined through computer simulation, which is the subject of the next chapter. The mathematical model can only be used to establish target points for the WIPAC Curve.

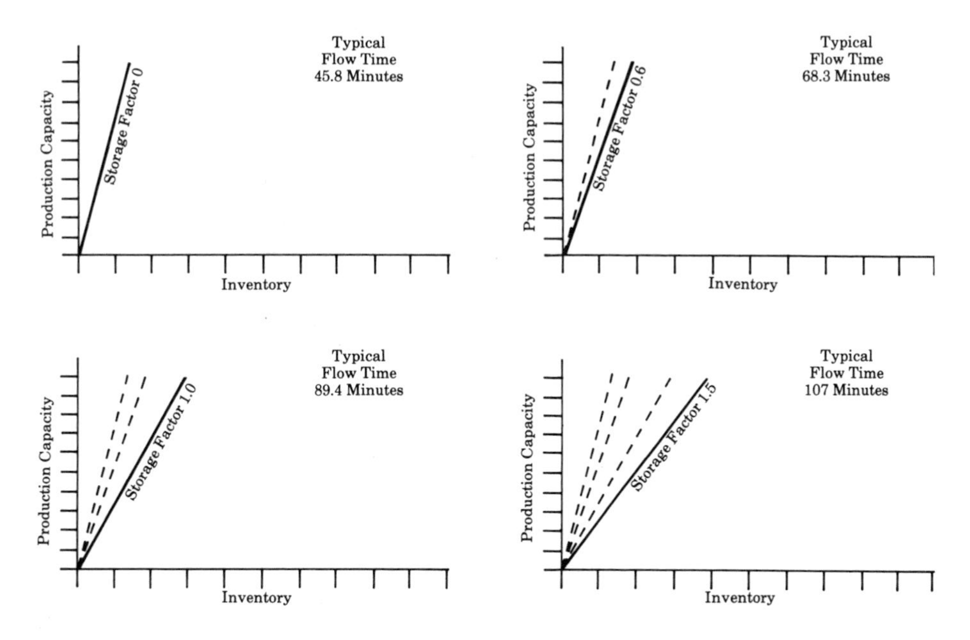

FIGURE 6-20 WIPAC graph with different storage factors.

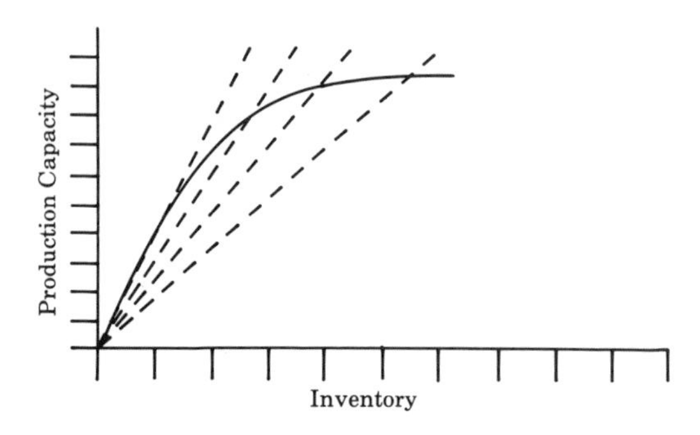

FIGURE 6-21 WIPAC curve intersection with various storage factor radians.

Figure 6-20 shows a series of radians, each representing the relationship between work in-process and system use. Each radian is computed from a fixed storage factor which yields specific flow times. Because these flow times remain constant, the WIPAC Curve is represented by a straight line. Therefore, a series of different storage factors are selected which yield different flow times. These then give different radians. The actual WIPAC Curve cuts through each radian at some level of station usage (Fig. 6-21). Where this curve cuts through is found through computer simulation. Chapter 7 contains the procedure for identification of this curve.

6.9 CONCLUSION

The aggregate planning of flexible manufacturing has the primary objective of establishing a feasible number of stations, transporters in-process storage capacity, and a work in-process level. Its main role in the design problem is to separate the decisions of capacity and integration. Computer simulation has been used to solve both of these problems, but sometimes an integration problem will appear as a capacity problem. By using aggregate planning, realistic targets can be established for system utilization and computer simulation can be used to feedback actual values for those which have been estimated, thereby focusing on the integration effects within flexible manufacturing

The design of flexible manufacturing is a most complex design problem. Without decomposing this into capacity, integration and control decisions, the proposed performance of a facility can be far removed from that which is actually observed during operation. The avoidance of this lack of relationship between computer models and reality is of top priority in the design. But, mathematical modeling is based upon assumptions and estimated data. When these are in any way uncertain, there is no substitute for the simulated operation of flexible manufacturing as part of its design.

7

Computer Simulation of Flexible Manufacturing

The most important yet most complex area for flexible manufacturing design is the study of the integration of a material handling system with a storage facility and work stations. A mathematical model can account for some aspects of this integration but the detailed study of component interactions can only be studied from the operation of the flexible manufacturing facility.

Two techniques exist to study the integration of flexible manufacturing. First, a physical model, such as the one shown in Fig. 7-1, can be constructed using scaled down replicas of each component. This technique allows the hardware aspects of integration to be observed and points out what special considerations might be needed to obtain automated integration. The second technique for studying the operation of flexible manufacturing focuses on an aspect of integration equally as important as the first: performance integration.

Computer simulation of flexible manufacturing systems operation is the best tool for studying the effects of integration upon overall system performance. Each component in the production facility will have some individual performance capability, but when this is integrated into flexible manufacturing, the adjustments to this performance must be known. The only method for determining these adjustments is through simulation, and the use of a computer is the most efficient means to accomplish this replication of operation.

Aggregate planning using mathematical models for flexible manufacturing can account for individual performance characteristics of components. When these components are integrated in flexible manufacturing, these individual performance levels become targets. Can the components achieve individual levels of performance when integrated into a low inventory, unbalanced, flexible production facility? Answering this question is the scientific role of computer simulation in flexible manufacturing design.

FIGURE 7-1 Physical mode to study integration of FMS. (Courtesy of M. Diesenroth, Virginia Polytechnic Institute.)

7.1 CATEGORIES FOR COMPUTER SIMULATION

Computer simulation falls into two broad categories as it applies to flexible manufacturing: general purpose simulation languages and specific languages (Fig. 7-2). The general purpose language has the capability to simulate any type of operational problem such as drive-up banking systems, traffic systems, job shops or flexible manufacturing systems. With this broad range of applications comes the increased skills necessary to use such a language.

7.1.1 General Purpose Simulation Languages

The individual using a general purpose simulation language is known as the modeler. Appropriately titled, this individual must construct the model of the system which is to be evaluated. Construction of a model involves translating real hardware components into conceptual terminology. For example, a part flowing through a job shop might need to be translated into a transaction moving through a network. Not only is it necessary to translate the system into

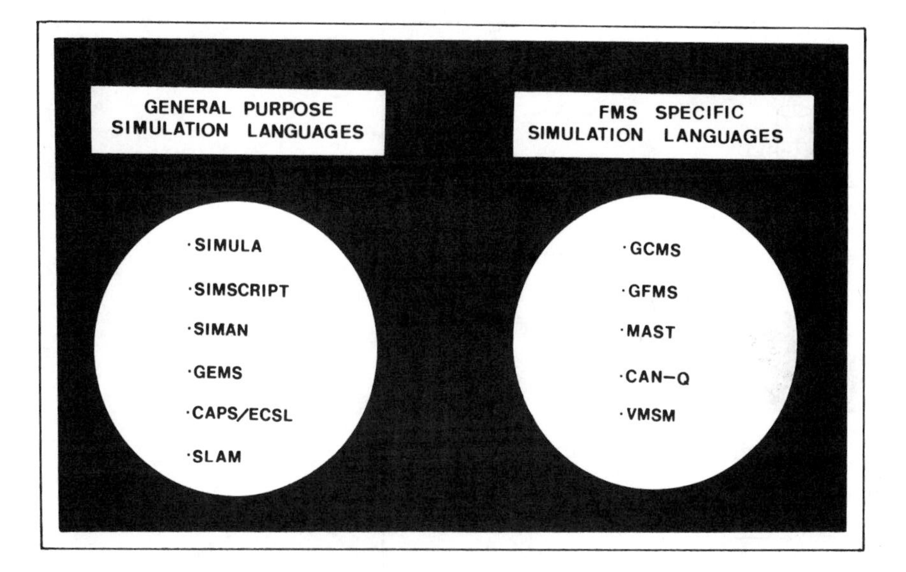

FIGURE 7-2 General purpose simulation languages and FMS specific languages. Shannon, R. E., "Intelligent Simulation Environments," January 1986, pp. 150-156.

generic terminology, but the logic of the operation must also be incorporated into the simulation model. The constraints of each component must be accurately accounted for or the simulation model will not be valid to the system under study.

In applying a general purpose language to the study of flexible manufacturing operation, the modeler must first translate physical components such as parts, carts, stations and pallets into generic terminology such as transactions, queues, network nodes and resource facilities (Fig. 7-3). Only when proper terms have been given can construction of the model begin. Model construction requires accounting for all constraints in storage capacity, cart traffic movements and station tool capacities. If any error or oversight is made, the results of the simulation will not reflect the real system performance. This validation of the model is the most important step in using a general purpose language and only experience and training can ensure that such problems are eliminated. Some studies have indicated that more than 1000 hours of education and use are necessary to obtain the skills required to use general purpose simulation to solve manufacturing problems.

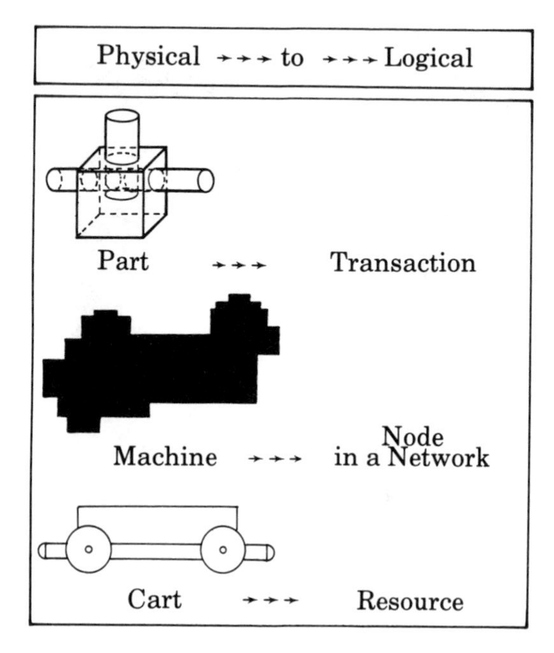

FIGURE 7-3 Model building.

7.1.2 Flexible Manufacturing Specific Simulation Languages

The second type of simulation languages are those which are specifically designed for flexible manufacturing simulation. The primary difference between the general purpose and the manufacturing specific simulation language in that the specific language will contain a validated model for manufacturing. Unfortunately, this model restricts the application of such a language to manufacturing type problems. But the advantage is that it is not necessary to construct a model and only data need be supplied in order to start a simulation. The education and training needed to use this type of simulation language will be a few hundred hours or less.

The model included in the flexible manufacturing specific languages will allow specific terminology between the real and simulated systems to be studied (Fig. 7-4). For example, parts are called parts and pallets are called pallets. A generic name, such as "station," might be used to represent a machine, robot, manual station, inspection or wash facility. Because each of these real facilities, designed to perform an operation, have similar effects upon the system, there is no loss of accuracy by the model in treating each as a generic station. The unique characteristics can

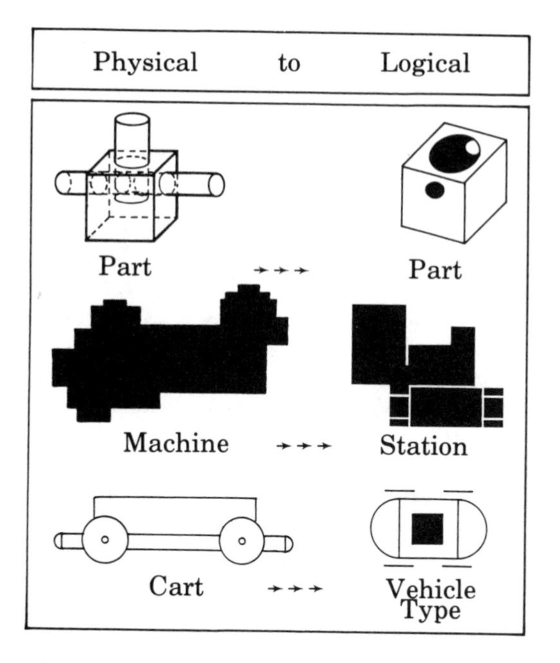

FIGURE 7-4 FMS specific model.

be accounted for by operation duration and types of parts which are allowed to be routed to this station.

Tooling characteristics in each station are accounted for by restricting the number of parts which can travel to specific stations. If a station has sufficient tooling to perform an operation, it will appear on a list of feasible stations for any part which requires this operation. Similarly, if a station does not contain sufficient tooling, it must not appear on the list of feasible stations. This approach to tooling constraints is acceptable only when tools are not automatically interchanged.

In batch production of a variety of part families, stations will frequently have their tools replaced with different sets to accommodate the change in production. When this occurs automatically, there is a need to simulate the changing over of the tools in a station. This means that as tools are picked up and delivered, the feasible set of stations which perform operations in the simulation model must change. The most accurate means by which to include this in the simulation model is to track each individual tool in each station. To determine if a station is eligible to perform an operation, the list of tools required for an operation must be compared to those tools which appear in the station at the time of station selection. This is not an impossible task but does require a huge amount of data.

This data must include a list of the tools needed for each operation, along with the usage, life and number of pots needed in the tool changer. Once this information is available, a feasible tool configuration can be assigned to each station. This allows some control to be maintained for these stations so that they will have tooling to perform specific operations. If a feasible tool configuration is not assigned, the tooling in the system may not reflect production needs and the system will become a study of massive tool changes at machines with very little production. As discussed in Chapter 6, feasible tool configurations can be computed from mathematical models without the need for complex simulation. Computer simulation of tooling should only be used to study the effect of changing over tools between tool configuration to meet planned changes in production.

7.1.3 Model Detail

The detail of the study of tooling in a simulation model is an example of the various breadths a computer model can accommodate. In some models, tooling is represented as the feasibility of the station with respect to given operations. In an application of a general purpose language to an FMS, each individual tool in the system is tracked and recorded each time it is used. Sometimes the detail of tracking individual tools is incorporated into the manufacturing specific simulation because the system may be used for the manufacturing of prototype parts where tool requirements vary between each part. In such systems, the effectiveness of the flexible manufacturing depends greatly on the ability to schedule tool changes which coincide with production schedules. The control and coordination in this situation needs to be studied and this has best been accomplished through the use of general purpose simulation with a specific model of the tool handling system proposed.

The study of the tooling is one example of the various levels of detail which can be incorporated into the simulation model key. The general purpose languages allow as much detail as is desired or as time permits, whereas the detail in the flexible manufacturing specific will be determined by the model which has been incorporated. In general, the flexible manufacturing specific language will have greater detail in material handling but may not be able to account for the system's unique tooling or fixturing aspects.

7.1.4 Simulation Output

The general simulation language differs in detail from the flexible manufacturing specific in the model, as do the types of summary statistics. In the general purpose applications, both the subject under study and the statistical results must be translated into model terminology. This translation is not a difficult problem to overcome for the experienced modeler but can be an obstacle for new users of general purpose languages.

Another important consideration between general purpose and flexible manufacturing specific simulation languages is the types of statistics which are available. In the case of flexible manufacturing specific, the choice of statistics has been decided by the model but these are mainly tailored to the information needs of the flexible manufacturing designer. The modeler in the general purpose language is able to select and even collect performance statistics, usually including the typical statistics on queue lengths and utilizations. Following is a list of the many performance statistics which have application in the design of flexible manufacturing.

7.2 INTEGRATION PERFORMANCE ANALYSIS

The role of computer simulation in flexible manufacturing design is to study the integration aspects of the system. This requires simulation of each part motion along with individual cart motion in the system, using data which is as accurate as possible. Because the study focuses on integration rather than capacity, outputs of performance, statistics from the simulation must include greater depth than simply queue lengths and utilizations. The description of the performance statistics for simulation is categorized into four areas.

7.2.1 Part Performance

Part performance includes production statistics, time in system and a breakdown of where a part spends its time while in the system. The production statistics, such as those seen in Fig. 7-5, will list all parts with the required production amounts. These amounts are represented by the level of production for each part, although the time period for this amount does not necessarily reflect the time period of the simulation. This makes it possible to use yearly production amounts and simulate the system for a shorter time period.

Along with the required production amounts will appear the actual number of parts scheduled and completed during the simulation

	PRODUCTION				FLOW TIME				
PART ID	PARTS REQRD	PARTS COMPL	PCT	AVG PARTS PER HOUR	AVG	STD DEV	MIN	MAX.	PARTS SCHEDULED
24	96	30	31	0.10	493.9	31.9	457.8	518.3	40
25	96	20	21	0.07	118.4	4.1	115.5	121.3	30
51	24	16	67	0.10	226.1	35.3	204.1	266.8	18
52	24	10	40	0.03	143.2	0.0	143.2	143.2	20

FIGURE 7-5 Part performance summary.

PART ID	AVERAGE TIME SPENT IN STATION TABLE	IN-PROCESS STORAGE	MATERIAL HANDLING
24	89.69	9.86	0.46
25	39.70	57.52	2.78
51	81.83	17.03	1.14
52	62.13	35.93	1.94

FIGURE 7-6 Performance percentage for completed parts.

period. Often, it is useful to calculate whether the desired production rate is being met by comparing a percentage of parts completed to a percentage required. This can be applied by using monthly production requirements and simulating one-fourth of this period. The resulting percentage of completed production must be at least 25% if the production rate is being met.

Another form of production output is use of the parts per hour production rate. This statistic is useful because most flexible manufacturing quotations include a guarantee of a production rate rather than a number of total parts. This will apply only when a sample of the parts, which will eventually run in the facility, are used for the simulation study. In this case, each part in the simulation might represent a family of similar parts where actual numbers of parts produced are not as meaningful as a production rate for evaluation of the total system performance. After the production statistics are presented, factors which provide additional data as to why the production rate or amount was at the observed level must be provided. This additional data provides insight into the previous results and establishes confidence in the accuracy of the results. One such support statistic is flow time.

The time in system statistic reports the average time which has elapsed from the time the part was first introduced until it was completed (Fig. 7-6). This time includes all operation durations, transportation and waiting time while in the system. In the discussion of aggregate planning, the flow time is identical to the

STATION TYPE NO.	METAL REMOVAL		ON—OFF SHUTTLE		PART STORED ON WORK TABLE		IDLE TIME	PCT	DOWN TIME	PCT
	TIME	PCT	TIME	PCT	TIME	PCT				
MILL 1	760.00	41.7	135.00	7.40	432.40	23.70	494.60	27.10	0.0	0.0
MILL 2	768.0	42.2	129.00	7.10	57.60	3.20	867.40	47.60	0.0	0.0
MILL 3	445.00	24.4	77.00	4.20	290.10	15.90	1009.90	55.40	0.0	0.0
GROUP AVERAGE	657.66	36.1	113.66	6.23	260.03	14.26	790.63	43.36	0.0	0.0

FIGURE 7-7 Station performance summary.

cycle definition. In reporting the time in system, the average overall completed parts are computed and the range of values is also notated. This range is most commonly reported as a minimum and maximum value with the standard deviation. When the standard deviation is no more than 15% of the average time in system, the system is producing parts at a predictable level. When this ratio of standard deviation exceeds 15%, there is a lack of performance control in the system. Following the statistic description is a discussion of integration analyses that includes more information on standard deviation levels and probable causes regarding loss of control.

To provide greater detail of the time in system, a breakdown of where the part spent its time while in the system is required. Simulation results will report the flow time as the sum of time while in operation, time in transportation and time in storage. The time in operation includes the total amount of time a part spends at all stations while positioned in the work table. This time will be at a minimum with the sum of all operation times in the part's route. This time can only be longer when a part has finished its operation but is unable to leave the work table because of transportation delays or in-process storage occupation. In any case, the average time in operation should be compared to the minimum in order to indicate how frequently parts are delayed at work tables. The frequency of such occurrences are also part of the station performance statistics described below.

The transport time represents the amount of time spent by a part on a vehicle or conveyor, either active or inactive. The in-process storage time is the elapsed time between leaving the work table and entering the transportation system, and leaving the transportation system and entering the work table. Specific design guidelines are available which provide strict levels for this storage time relative to station time. These guidelines are discussed in the integration analysis section.

7.2.2 Station Performance

The station performance summary in Fig. 7-7 includes statistics for each station and the percentage of time each station is occupied by various states of activity. These states include being busy, idle, blocked and down but can result from a variety of conditions. For example, a station can be busy while performing an operation or part motion of loading or unloading its work table. Occasionally when all the storage positions are occupied, a work table becomes blocked because the finished part has no place to go. Some definitions classify this condition as part of the station's idle time due to its lack of activity. Regardless of which method is used in reporting this condition, it reduces station capacity and the statistics must reflect this.

It appears that the clearest means by which to report these states of activity is through the use of separate statistics. This includes reporting the total times spent in operation, shuttling a part to or from the work table, idle with a part blocked on the work table, idle with no part on the work table and the total time down for each station. Accompanying the total time will be the relative percentages of time during the simulation period for which the station was in each of these five states.

These statistics must be reported for each separate station and should be averaged for all station groups. A station group is a collection of individual stations which can perform all of the same operations. Stations which are members of the same group are identically tooled, making specific station operation selection unnecessary. Once the stations meet this criteria for members of the same group, their individual statistics should be averaged.

This averaging eliminates the distraction of one station in the group having different individual statistics from another. As far as integration problems are concerned, this group will have a target level of utilization and the simulation should report its results accordingly. The disparity between individual station statistics which are members of the same group is a control problem in that the algorithm for selecting a station might be giving bias to a subset of the stations. The treatment of such problems is part of the control evaluation step in flexible manufacturing systems design and is described in detail in Chapter 8.

7.2.3 Transporter Performance

In Fig. 7-8, the transporter performance summary includes statistics on the five possible states of a transporter. A transporter can be moving, shuttling a part to or from a station, stopped with an assignment while waiting for a path to become free, stopped without an assignment and down. The total minutes should be reported along with the relative percentage of time spent in each position.

As in the case with station groupings, transporters which service the same area and lie in the same track can be replacements for one another. In these cases, average statistics should be reported for all individual transporters in the group. This provides an observed value of utilization which can be compared to analytic calculations. The discrepancy between individual transporter utilizations is not an integration problem but rather a control problem. These problems are best solved using animation, which is discussed in the next chapter.

	CART ACTIVITIES											
CART TYPE NO.	MOVE TIME	PCT	SHUTTLE TIME	PCT	DOWN TIME	PCT	IDLE TIME	PCT	TOTAL DISTANCE MOVED	NO. OF ASSIGN-MENTS	AVG. MOVE DISTANCE	AVG. MOVE TIME
AVG 1	505.8	36.20	583.98	41.7	0.0	0.00	310.22	22.1	95686.0	285	335.74	1.7
AVG 2	413.0	29.50	486.10	34.7	105.2	7.50	395.70	28.3	80373.0	191	420.80	2.1
AVG.	459.4	32.85	535.04	38.2	52.6	3.75	352.96	25.2	88029.5	238	378.27	1.9

FIGURE 7-8 Cart performance summary.

7.2.4 Inventory Performance

Pallet performance, as shown in Fig. 7-9, includes statistics on the average number of parts in process, and the minimum and maximum number required during the simulation. The maximum number used is the largest quantity of parts which exist simultaneously in the system. The minimum number is the smallest amount, and the average is computed from the relative time each quantity of parts is in the system. The accumulation of all average number of parts will yield an average work in-process level used in the system. The average number of parts can be summed for all types of parts to give an average work in-process level.

With the mathematical models described in Chapter 6, the assumption was to produce all parts continuously throughout the production period. Often, however, the flexible manufacturing provides for the manufacture of a variety of parts and continuous production is not feasible. Instead, subsets of these parts can be batched, establishing the number of pallets when production rates need to be higher for short periods of time.

Computer simulation can be performed with batch type scheduling. Typical batch varieties and quantities can be introduced into the system with the collection of pallet statistics. These statistics will provide the average and maximum numbers used during the batch period. One characteristic of flexible manufacturing operation is that the work in-process level will remain approximately the same during either batching or continuous production. These characteristics and others are discussed in the integration analyses for flexible manufacturing.

PART TYPE NO.	IN USE AVG.	IN USE NOW	NOW AVAILABLE	MAX. USED
1	2.0	2	0	2
2	1.2	1	1	2
3	0.8	1	0	1
4	1.3	1	2	3
5	0.3	1	1	1

FIGURE 7-9 Pallet performance summary.

7.3 FMS PERFORMANCE ANALYZER

The design guidelines which were discussed as part of aggregate planning resulted with target values for station utilization, transporter utilization and flow time for each type of part. These targets provide a measure from which the simulation results can be compared. By using these mathematical results as a base, the effects due to integration within flexible manufacturing can be isolated from the capacity of each component.

This decomposition of integration effects from capacity is the primary role of computer simulation. Computer simulation can simulate the proposed operation of the facility. During the simulation, performance statistics can be collected and reported. These results, which include the integration effects, need to be compared to the mathematical results in order to isolate the actual integration effects. Following is a procedure for making this comparison.

The procedure starts with the normalization of the mathematical results. This step will align the mathematical results with the simulation results. Once the results are aligned, the comparison can be made between the simulation and mathematical results. A step by step description of the analysis of integration effects follows and is seen in Fig. 7-10.

7.3.1 Step 1: Construct the WIPAC Curve

The first step in identifying the integration effects is to quantify the relationship between inventory level, flow time and production.

1. Construct WIPAC Curve
   - A. Prepare data and run simulation
   - B. Compare mathematical results with simulation results
   - C. Adjust inventory level to align mathematical results with simulation results.
   - D. Repeat steps A—C until convergence
2. Test for uniform resource allocation
3. Quantify the integration effects

FIGURE 7-10 Expert analysis.

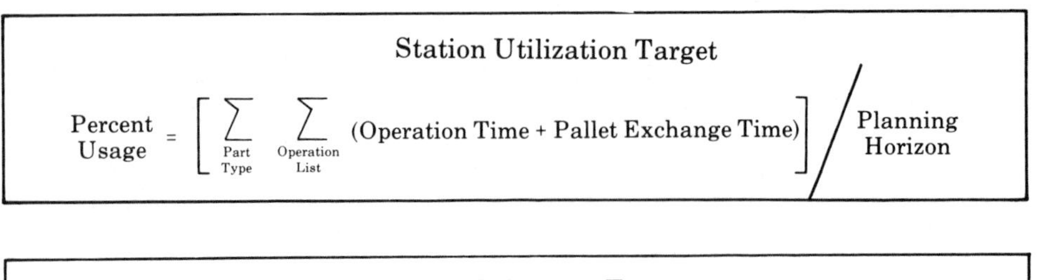

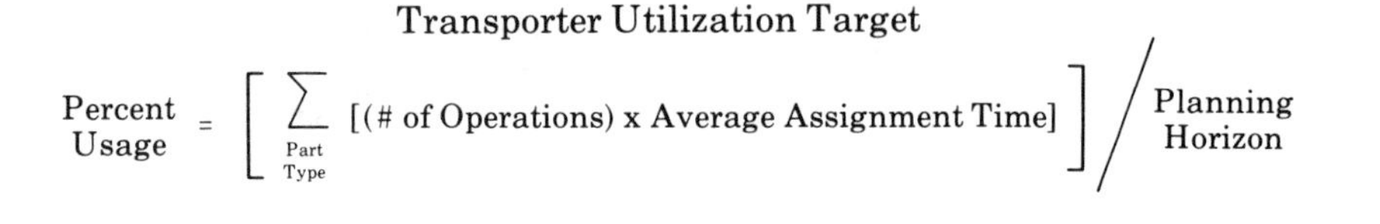

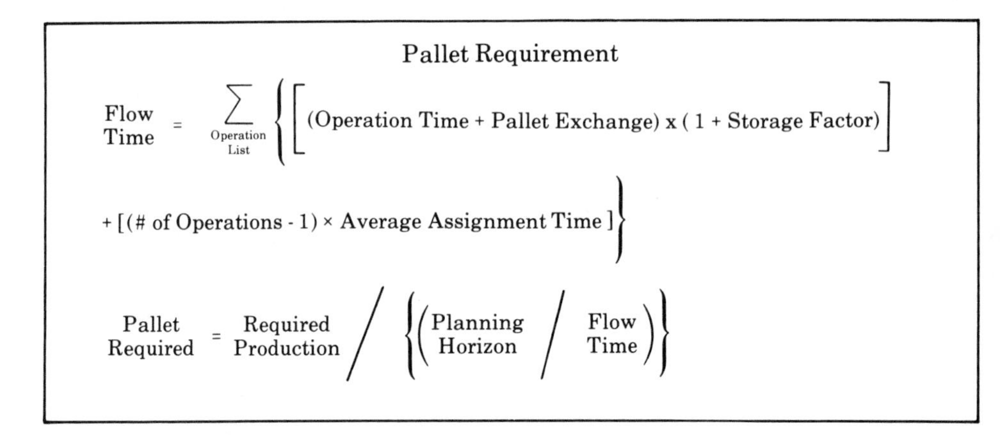

FIGURE 7-11

These variables are displayed through the WIPAC Curve. Therefore, the first step is to generate the WIPAC Curve.

The procedure utilizes the mathematical model to set initial conditions for the simulation. From the required production amounts and process plans, the target level of station utilization can be determined, as calculated in the mathematical equation in Fig. 7-11. With an estimate of the average transporter assignment time, the number of vehicles and the target utilization can be determined. With an estimate of the storage time for a part, a target for its flow time can be determined and from this result, the appropriate inventory level determined.

The generation of the WIPAC Curve is an iterative process where the estimates in the mathematical model are adjusted according to the observed results from the simulation. For this reason, the effects of the estimate are eventually filtered out of the mathematical model and only serve as a means to set the initial condition for the simulation.

7.3.2 Starting the Iterative Process

To start the iterative process for the generation of the WIPAC Curve, an initial guess is needed for the storage factor and transport assignment time. The transport assignment time can be estimated from the elapsed activity time for move to pickup position, load a part, move to deposit position and drop off a part. Estimating the storage factor for each part is less scientific and uses a rule of thumb instead. The rule of thumb for guessing the initial estimate for the storage factor is to use a value of zero for all parts. This value ensures that the initial conditions for the simulation will yield a point on the upward portion of the WIPAC Curve. This is desirable because convergence tests are made simpler and the interesting portion of the WIPAC Curve will be generated. Figure 7-12 shows the first target point for the WIPAC Curve generation. From these initial conditions, the simulation will produce an actual point on the WIPAC Curve.

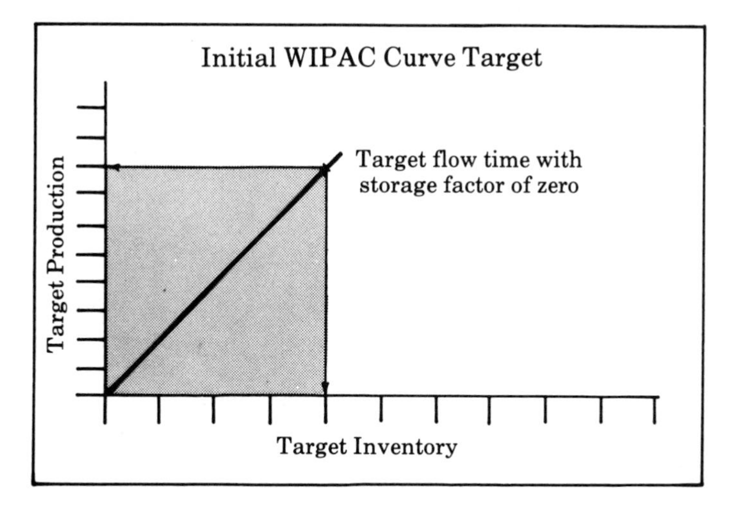

FIGURE 7-12

7.3.3 Simulation Results

Upon completion of the simulation, the results provide actual observation of inventory, flow time and production rates. The inventory is reported in terms of average parts in process. The flow time is the average throughput time of each part and the production can be measured by either a production rate or station utilization (station utilization statistics will be most useful when a clear definite bottleneck station exists). This data provides an actual point on the WIPAC Curve as shown in Fig. 7-13. Upon completion of the simulation, an actual point on the WIPAC Curve is known along with the target point from the mathematical calculation. The comparison of these two points do not yield meaningful information because the target is based upon "best guess" data. These guesses bias the mathematical results and, at this time, might not even be feasible characteristics for this manufacturing system. Therefore, before the target points can be compared to the simulation results, the bias of the target part must be removed.

The bias in the mathematical calculation can be removed by using the observed values from the simulation as replacements for the "best guess" data. Such data includes the average assignment time and the storage factors for each part type. Therefore, these values are identified from the simulation results and inserted into the mathematical target equations.

This process results in a new target point. This point lies on a radian from the origin of the WIPAC Curve and passes through

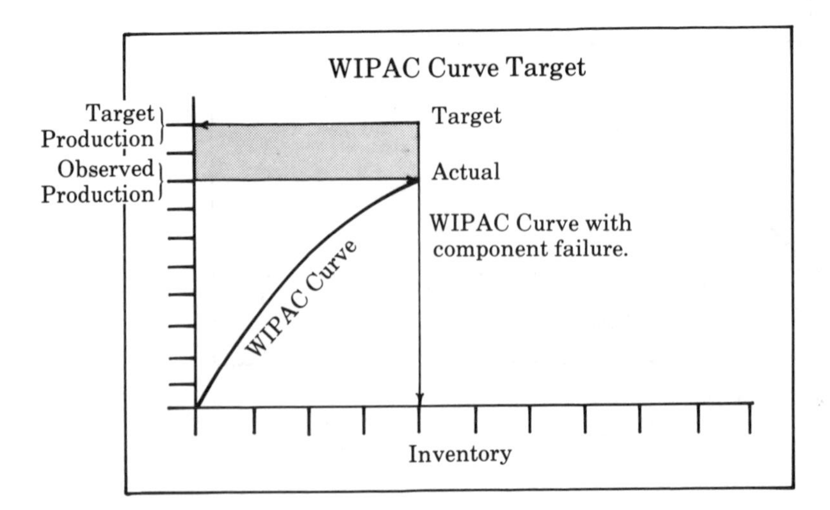

FIGURE 7-13

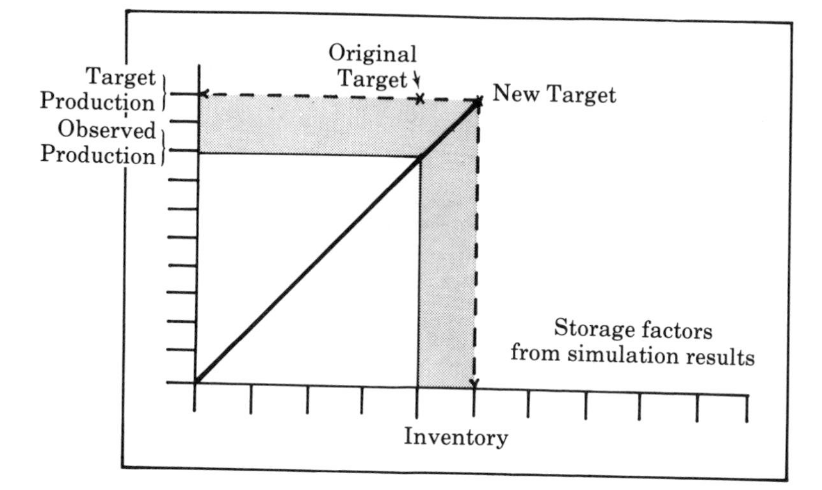

FIGURE 7-14 WIPAC Curve with aligned target.

the actual WIPAC Curve point identified with simulation results. Figure 7-14 shows the WIPAC Curve with the aligned mathematical target points which are used in the test for convergence. However, if convergence is not reached, the aligned target point is used to set the initial conditions for the next simulation run. The difference in the initial condition will be to increase the number of pallets to an appropriate amount which is consistent with the observed flow times of the parts.

7.3.4 Convergence Test

The actual point in the WIPAC Curve and the aligned mathematical target point are used in the test for convergence. The convergence test is used to determine when there is no need for further generation of the WIPAC Curve. There are three conditions where convergence has been reached. Following convergence, it is not necessary to perform further iterative generation of the WIPAC Curve.

Convergence Test #1 Convergence has been reached when the aligned target point matches that of the observed WIPAC Curve point (Fig. 7-15). When this condition exists, the simulation results will meet or exceed the target level of station utilization.

Convergence Test #2 Convergence has been reached if the WIPAC Curve bends downward (Fig. 7-16). This can be identified when the production rate or station use drops on consecutive simulation

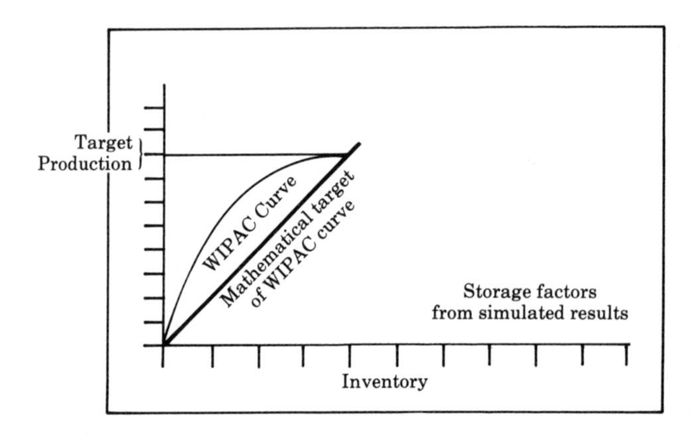

FIGURE 7-15 WIPAC Curve graph test #1.

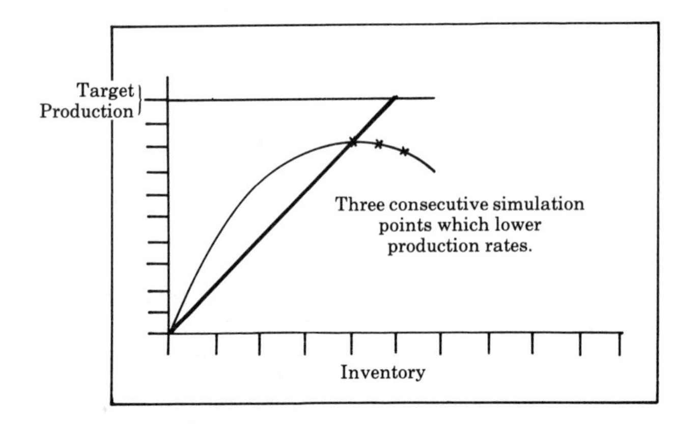

FIGURE 7-16 WIPAC Curve graph test #2.

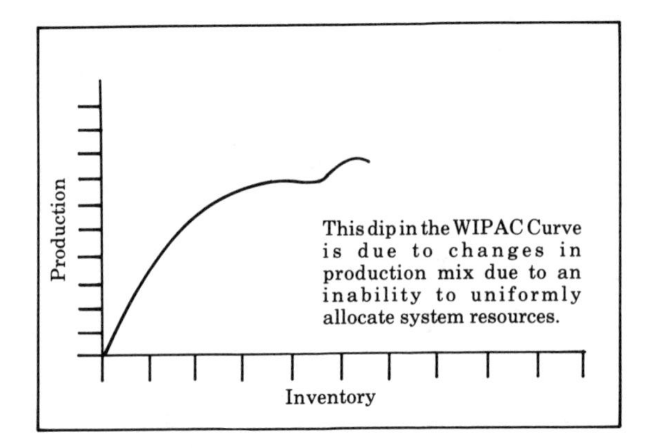

FIGURE 7-17 WIPAC Curve graph test #2 with production changes.

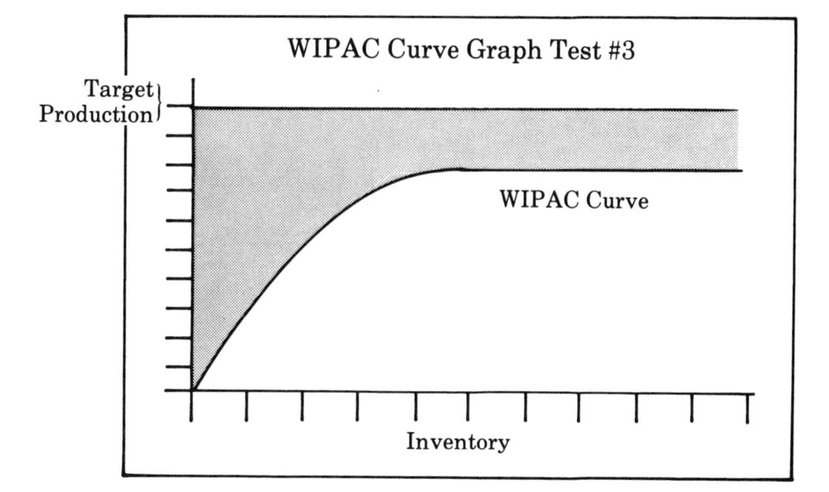

FIGURE 7-18

runs. Caution is advised because changes in the production mix will cause dips in the WIPAC Curve (Fig. 7-17). Convergence is reached when there is a true trend of downward movement observed in the WIPAC Curve.

Convergence Test #3 Convergence has been reached if the WIPAC Curve becomes flat. Figure 7-18 gives an example of this situation which has been detected from results of constructive simulation runs. This flattening out of the WIPAC Curve is due to the existence of a clear bottleneck that might be found in the transporter system or result from a lack of storage space for in-process parts. The identification of the cause is part of step #3 in identifying integration effects.

The iterative process of using observed simulation results in replace estimates in the mathematical model, which in turn set the initial conditions for the next simulation.

From this technique, the variation in the initial conditions will be the inventory level. There are two points of caution for ensuring the convergence of the WIPAC Curve has actually been reached. First of all, the number of parts set in the initial condition of the simulation must be an integer where the mathematical model might arrive at a portion of a part. When the simulation results are close to convergence, the change to the storage factor might not result in a change to the integer number of parts. When this occurs, an arbitrary adjustment to the number of parts is needed. In a sense, a small jump to the next integer is needed to avoid the integer solution problem.

The second point of caution is that the mathematical model will have a target level of production. Due to the mathematics, the initial conditions for a simulation cannot be generated for levels of production which are higher than the target. Therefore, some temporary adjustments to the planning horizon might be needed to reach convergence for the WIPAC Curve.

When convergence has been reached, the highest point of the WIPAC Curve is used to evaluate the overall productivity of the manufacturing system.

7.3.5 Step 2: Test for Uniform Resource Allocation

The allocation of resources within the manufacturing system will not be uniform to all parts. Some parts will get more than their share and others will get less, which will be observed in variations of production rates (Fig. 7-19). These nonuniformities of the manufacturing system operation need to be accounted for before starting the evaluation process. If they are not accounted for, the result of the simulation can show that station utilization has exceeded the target but the desired production will be observed through changes in production rates of the part types. As these relative production rates vary, this will cause deviations along the WIPAC Curve.

These deviations will automatically be accounted for in the use of the mathematical model to set the initial conditions for the next

PART TYPE	REQUIRED PRODUCTION RATE	OBSERVED PRODUCTION RATE
A	2.0/HOUR	1.5/HOUR
B	3.1/HOUR	4.7/HOUR
C	7.3/HOUR	7.0/HOUR
TOTAL	12.4/HOUR	13.2/HOUR

FIGURE 7-19 Required vs. observed production rates.

simulation. The only point of concern is at the highest point of the WIPAC Curve, because this is the point which will be used to measure the productivity of the manufacturing system.

The method for testing the uniformity of resource allocations within the manufacturing system is to use the production rates of the individual part types. A target production rate for each type of part is computed from the mathematical product and these are compared to the observed results from the simulation.

The deviation in production rate is the difference between the target rate and the observed rate. These deviations are computed for each type of part. When these deviations vary in value, the allocation of resources in the simulation is different from that which is assumed in the mathematical model. If these vary by more than an acceptable level, the manufacturing system has not allocated its resources in an acceptable way. When this conclusion is reached, there is no need to study the productivity and action should be taken to correct for this imbalance.

The direct approach to eliminating the imbalance of resource allocation is to introduce greater flexibility in the manufacturing facility. There is a cost for this additional flexibility but there is also a cost for the lost productivity in flexible manufacturing due to its inability to allocate resources at a desired level.

7.3.6 Identify the Integration Effects

Using the equation for net production described in Chapter 2 the integration effects can be measured as the difference between the gross production capacity and the net capacity. The gross production capacity is that established in the mathematical model and the net production is observed in the simulation results.

When net production equals gross capacity, the FMS has no lost capacity due to integration effects. Figure 7-20 illustrates this condition in the WIPAC Curve. In this case, the appropriate inventory is identified, flow times are known and the FMS can realistically achieve this level of productivity. Now the study repeats with some interruptions into the station availability. The following section contains a description of this procedure.

When net production is less than the gross target, integration effects are present in the flexible manufacturing system. Figure 7-21 illustrates this condition in the WIPAC Curve. These integration effects are a function of inventory, balance and flexibility. Further analyses is to identify the marginal impact each of these have upon the integration effects. One consideration is what increase in in-process storage capacity would be needed to reduce integration effects. But increases in inventory is only one of three contributing characteristics to integration effects.

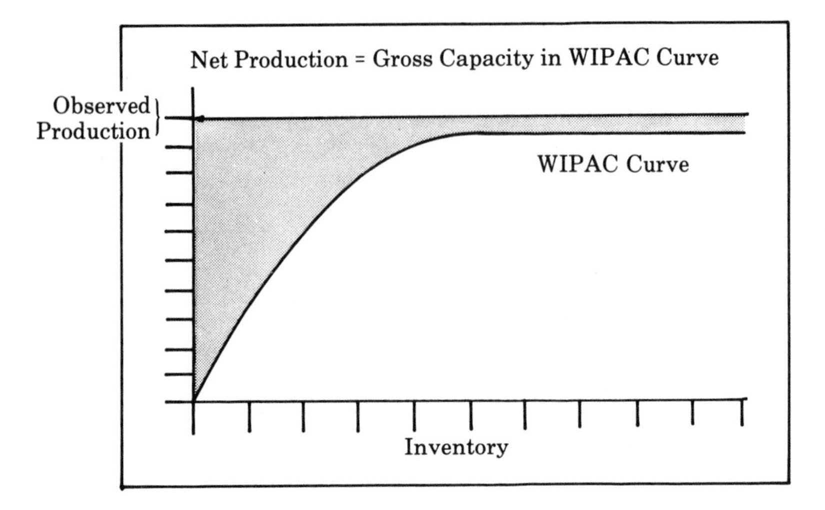

FIGURE 7-20 Net production = gross capacity in WIPAC Curve.

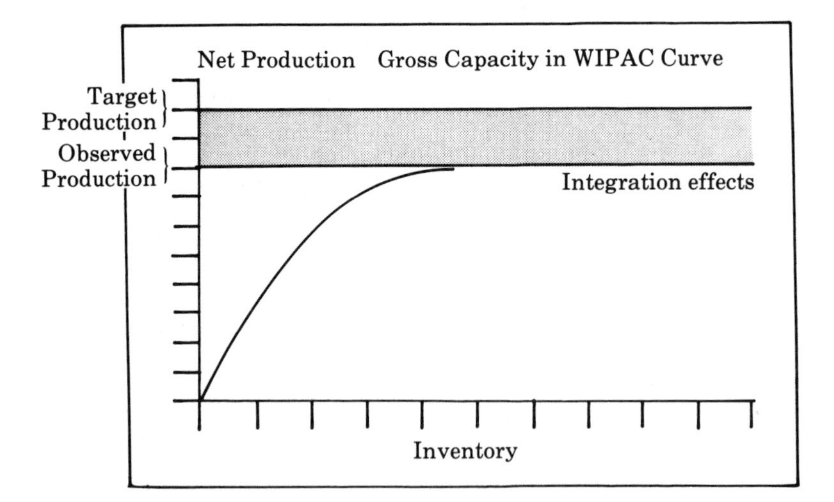

FIGURE 7-21 Net production < gross capacity in WIPAC Curve.

1. Is target transporter use equal to observed value?
2. Where are pallets used?
3. Are any operation durations close in value to average transport time?
4. Are observed use of individual stations within a group uniform?
5. Did the simulation run long enough?
6. Did the sample reach steady station operation?
7. What is the adjust system capacity when component failure is included?

FIGURE 7-22 Rules for determining integration effects.

Flexibility is a second candidate for causing these integration effects; however, when the system allocates resources as desired, its impact upon integration effects is reduced. The dominant cause for integration effects at this point is the balance within the flexible manufacturing system. In this sense, balance refers to a general relationship between the utilization and cycle times of all system components. This includes parts, transporters, work stations and storage capacities. Determining which of these components or combination of components is creating the bottleneck requires a set of rules. A tentative list of these rules is shown in Fig. 7-22 and provides a basis for implementation of an expert system. These rules are preliminary and provide a framework for expertise in flexible manufacturing evaluation to be implemented through an expert system. Research is continuing in this area.

7.3.7 Integration Effects with Station Failures

The above procedure is useful for identifying integration effects which are inherent in the design of flexible manufacturing. However, this analysis does not provide sufficient information for the integration effects where dynamic station availability occurs. The integration effects which occur when station failures and repair occur are the most important for determining the net or real production capacity of the FMS. When stations fail, the effect is to lower the gross production capacity by a direct amount. However, the dynamic interruption will cause carry-over effects due to material shortages and the extent of these effects will be determined by the

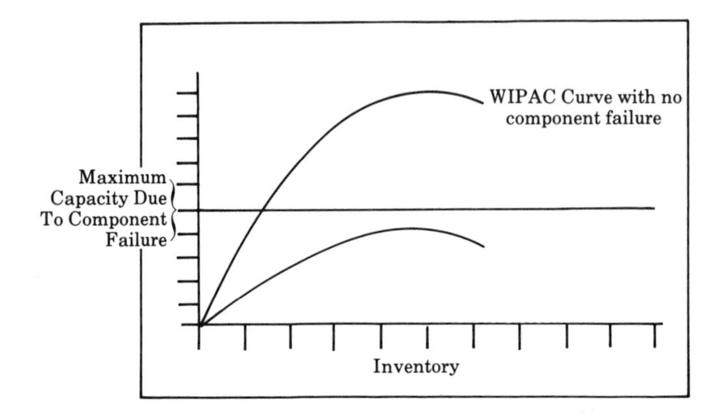

FIGURE 7-23 Station failures in the WIPAC Curve graph.

inventory, balance and flexibility in the manufacturing system. In most cases, the occurrence of station failures acts like a weight on the WIPAC Curve (Fig. 7-23). This weight not only pushes the curve down but also has a tendency to roll the curve forward. For this reason, the optimal point when no station fails could become a point on the downward sloping portion of the WIPAC Curve when a station fails. This phenomenon is the most common problem of flexible manufacturing today. Because of this, the integration effects through the generation of the WIPAC Curve are important for predicting the real productivity of the manufacturing system.

The procedure for identifying the integration effects when station failure occurs is identical to that described above. The only difference is that some planning time is taken away from the station in the mathematical model and some repair pattern is introduced into the simulation. For example, if one station is to lose 20% of its capacity, the load calculation for this station cannot exceed 80% of the planning horizon. This will establish a target production capacity. The simulation must then have data added which describes the time between failure and repair time probability distributions. These distributions must reflect the average lost capacity which has been included in the mathematical model. When this has been confirmed, the procedure for generation and evaluation of the WIPAC Curve is identical to that described above.

7.4 CONCLUSION

Computer simulation of flexible manufacturing has the primary objective of providing a scientific means by which to study the integration effects upon the various components. This is possible when the capacity decisions have been resolved using an analytic model and the decision of how many stations and transporters are necessary has been determined. Computer simulation produces an abundance of statistical results. These results must be reviewed in a hierarchical nature, each adding further insight into the performance of flexible manufacturing. This hierarchy of components should then be studied, which requires a knowledge of how this integration affects the statistics. However, once understood, specific problems can be isolated in a relatively short time with use of just a few statistics.

Even after the integration problems have been identified, not all can be resolved with simple reconfiguration of the system. These problems must be resolved with control strategies which maintain control over congestion and optimize the total system performance. The next chapter describes the study of these problems involving graphic animation of simulation results.

8

Control Evaluation and Automation

Computer simulation provides a means to study the effects of integrating islands of automation into a flexible manufacturing system. In this study the hardware integration effects are reported by use of performance statistics which can be compared to the results from analytic models of FMS. The comparison between simulation and mathematical results provides a direct method for analyzing component and integrated system capacity. However, the study of integration must not be limited to hardware considerations; it must also include the integration effects on the operation of the FMS.

8.1 FMS OPERATION HIERARCHY

Flexible manufacturing system's operation and its control coordinate the activity of all components with a primary objective of efficiency. This control is accomplished by the transmission of component status through a communication network (Fig. 8-1). Comparising a network are various computers which collect status information and store it in a database. Control algorithms react to this data and issue suitable commands. The process of collecting status information that results in commands of operation is known as real time control. Commands are referred to as real time decision making and the logic of each decision is accomplished by a control algorithm.

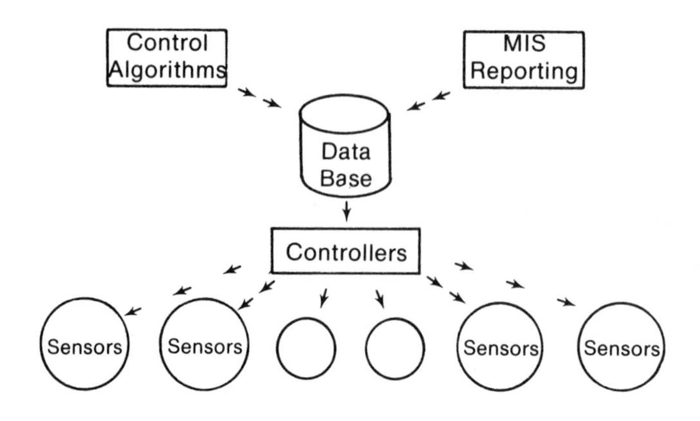

FIGURE 8-1 FMS operation and control hierarchy.

8.1.1 Control Algorithms

Control algorithms provide the guidelines by which the flexible manufacturing systems will operate. As important as hardware integration is, it is just as important for control algorithms to efficiently utilize this integration. If hardware is integrated with a capacity proven by mathematical results, control algorithms must be compatible with the capabilities and limitations to obtain optimal efficiency. Therefore, flexible manufacturing design must include a study of these controls to produce specific algorithms by which the system will operate.

8.1.2 Measuring the Effectiveness of Control Algorithms

Computer simulation replicates all of the motion within the FMS and in doing so uses algorithms to make alternative decisions. What must be determined is how to measure the effectiveness of these algorithms in controlling the operation of the system. Performance statistics, summarized in Fig. 8-2, yield whether the production rate or station utilization has reached targeted levels. But these statistics do not produce much information about the efficiency of individual decisions during the system operation. Greater detail is needed to the point where every action of the simulated FMS should be traced and reported.

The method of reporting detailed activity which takes place in a simulation is in the tracing of system status progression. This is commonly referred to as the event trace (Fig. 8-3). A typical event trace contains statements that describe each event through

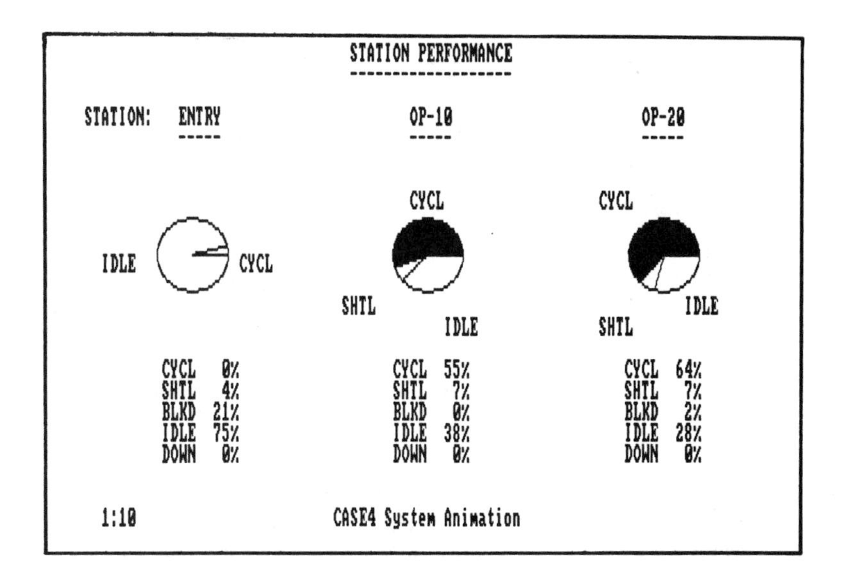

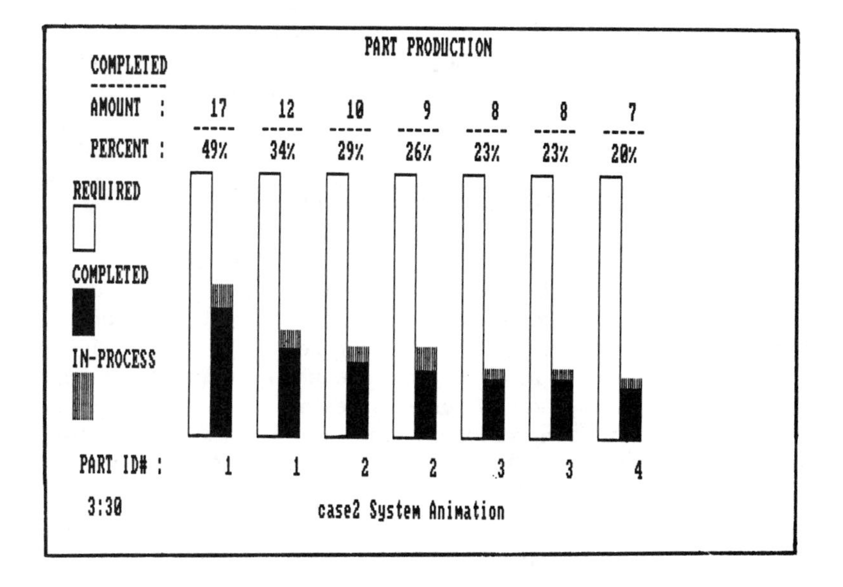

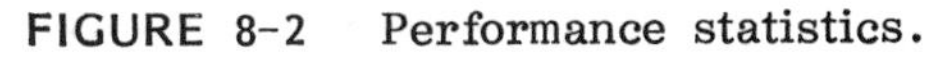
FIGURE 8-2 Performance statistics.

```
AT TIME   62.011   PART TYPE   1  COMPLETED OPERATION   4  ON STATION   7
AT TIME   62.551   PART TYPE   3  COMPLETED OPERATION   6  ON STATION   4
AT TIME   62.611   CART NUMBER   1  ARRIVES AT DECISION POINT   8
AT TIME   62.651   CART NUMBER   1  ARRIVES AT DECISION POINT   7
AT TIME   62.691   CART NUMBER   1  ARRIVES AT DECISION POINT   6
AT TIME   62.781   CART NUMBER   1  ARRIVES TO PICK UP PART TYPE   3  AT STATION   4
AT TIME   62.801   PART TYPE   3  MOVES INTO OFF-SHUTTLE OF STATION   4
AT TIME   63.051   PART TYPE   3  LOADED ON CART NUMBER   1
AT TIME   63.141   CART NUMBER   1  ARRIVES AT DECISION POINT   6
AT TIME   63.181   CART NUMBER   1  ARRIVES AT DECISION POINT   7
AT TIME   63.221   CART NUMBER   1  ARRIVES AT DECISION POINT   8
AT TIME   63.281   CART NUMBER   1  ARRIVES TO DROP OFF PART TYPE   3  AT STATION   7
AT TIME   63.331   PART TYPE   3  MOVES INTO ON-SHUTTLE OF STATION   7
AT TIME   63.381   PART TYPE   3  MOVES INTO MACHINE AT STATION   7  TO BEGIN OPERATION   5
AT TIME   66.381   PART TYPE   3  COMPLETED OPERATION   5  ON STATION   7
```

FIGURE 8-3 Event trace.

coded information, such as part number, in addition to an English description of each event and its corresponding simulation time. The event trace makes it possible to follow the transition from one system status to the next.

Because it is impossible to remember the status of all parts, stations and transporters, one cannot simply read through a list and obtain the information as to the efficiency of the decisions for given situations. Visualization of the status is essential to provide this memory.

8.2 VISUALIZATION OF SIMULATED ACTIVITY

One method for achieving visualization of simulation status is by using a sketch of the FMS layout for background (Fig. 8-4) and moving markers in the foreground as directed by the description contained in the event trace. Although this method is frequently used in verifying a simulation, it is time-consuming, and any inaccurate movement of the marker requires a restart of the tracing process. Therefore, it is necessary to provide an automated means for replaying the event trace to study the effectiveness of control algorithms in the FMS simulation.

8.2.1 Color Graphic Animation

Automation of the simulation event trace can be accomplished through color graphic animation. Color graphic animation involves recording images one frame at a time to create an illusion of motion. The most common method of creating this illusion involves blinking the same image in succession at different points in the computer display

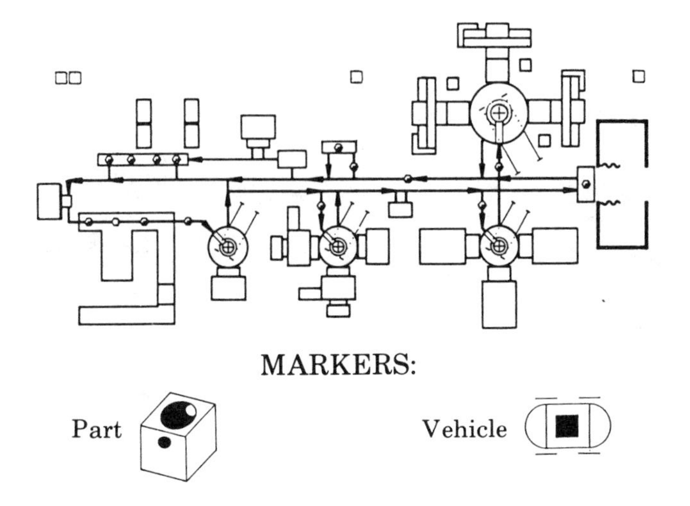

FIGURE 8-4 Visualization of simulated activity.

background. The illusion of motion is produced by moving the image slightly from its previous point of reference in the display. The smaller the displacements and higher the frequency of each, the more continuous the illusion of motion. But each frame must correspond to some status of the system, and providing continuous motion would require thousands of frames.

The application of animation to computer simulation output is accomplished by automation of the event trace. A background is drawn on the computer display which resembles the actual layout of the FMS under study (Fig. 8-5). Animation is achieved by processing the event trace containing the logical sequence of activities and appropriate time delays to blink corresponding images (Fig. 8-6) through the background. There are two methods for obtaining animation from computer simulation. The first is to synchronize the animation with the simulation. In this method, the animation display reflects the current status of the simulation. Here it is possible to interrupt the animation and inject an alteration to a component's status. Once this has been done, the simulation can resume from its current point with the altered status, permitting designer interaction during the course of the simulation. But these injections must represent realistic situations or the animation can become an expensive video game for the amusement of the designer. The alternative method for achieving animation from simulation is to run the animation as a postprocessor. In this method, the events are recorded in a data file as the simulation runs. This permits the

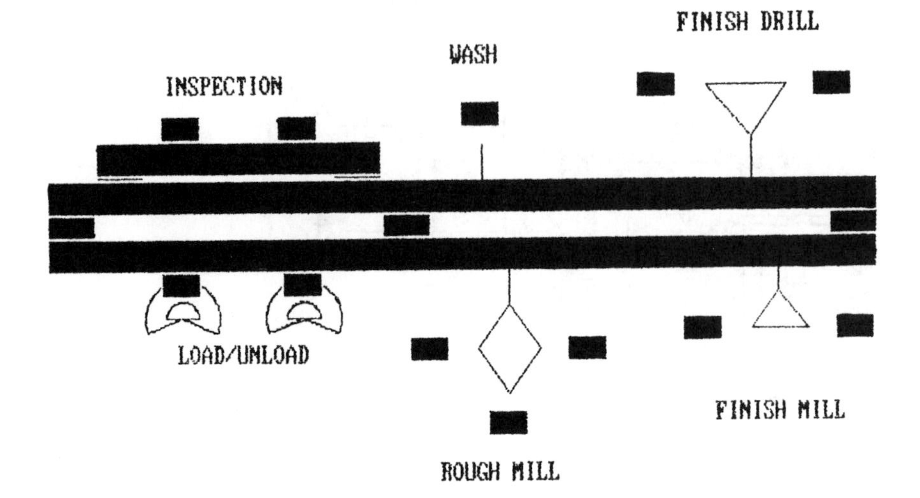

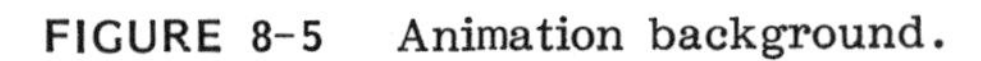
FIGURE 8-5 Animation background.

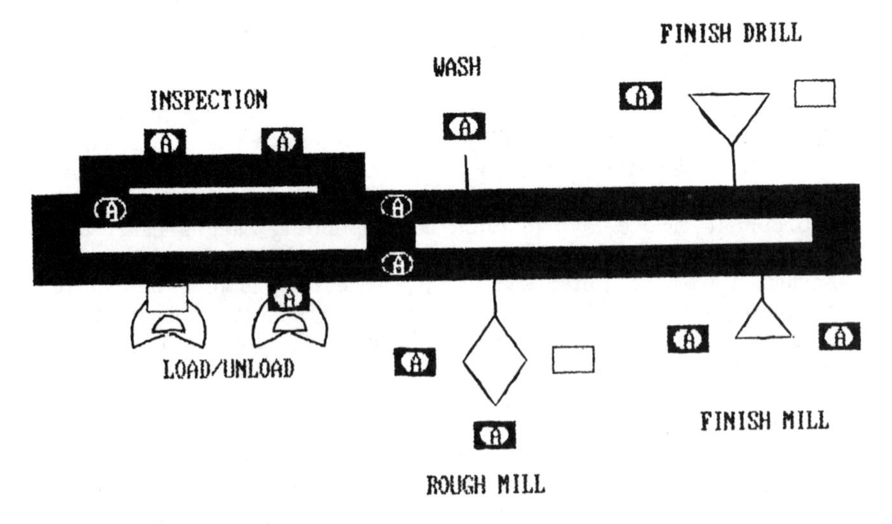

FIGURE 8-6 Corresponding images.

simulation to complete and produce its performance statistics before any animation can be viewed. Because the simulation is complete before the animation can be seen, designer interruptions with status change are not possible when the animation is not synchronized with the simulation. To study these injections of status change, they must be planned as part of the simulation parameters.

Because the performance statistics are available, they can be reviewed. From this analysis, specific problem areas can be identified and then serve as focal points for the animation. It is always easier to see the problem when we know where to look! Both methods, synchronized and nonsynchronized, have advamtages and disadvantages. The primary advantage for the user of synchronized animation is the ability to inject dynamic changes.

8.2.2 Selection of Images

One problem associated with animated simulation is the difficulty in selecting images and methods for representation of the various status of components. This is especially true for animation of general purpose simulations. Because these languages use generic terminology, a model must be constructed. However, the resulting animation becomes an animation of the model created and may not be visually similar to the actual system under study.

This problem is overcome in the application of languages intended for flexible manufacturing type studies. This is because the model in the simulation language will have a direct relationship to real world components. With this one to one relationship, it is possible to chose graphic images which visually resemble real FMS components (Fig. 8-6). For example, pallets can be represented by circles which contain some color or identification indicating if they are empty or the specific type of part fixtured. Carts can be represented as rectangles with stations drawn as boxes for the work table and in-process storage positions. The illusion of motion can be accomplished by blinking the pallet images through the carts, storage positions and work tables. When in the work table, color or text can be used to indicate the status as busy, blocked or down. The advantage of using images which have a direct visual relationship between their real world counterparts is that anyone with a brief knowledge of manufacturing can understand and follow all of the activity of the simulation. This feature in itself enables simulation to become more effective as part of the decision support system in manufacturing system evaluation.

8.2.3 Detail of System Layouts

The amount of detail in the background drawings is usually dictated by the graphic device capabilities. The inexpensive microcomputer

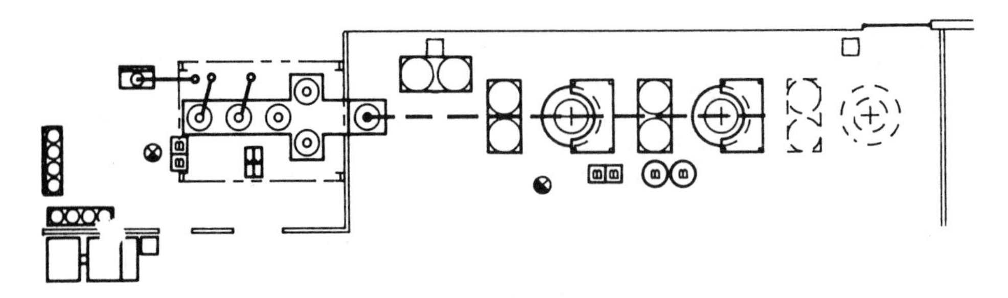

FIGURE 8-7 CAD drawing

will provide medium resolution with a pallet of four or sixteen colors. This will permit only simple images such as boxes, rectangles and circles. The other extreme is the use of a Computer Aided Design (CAD) graphics system (Fig. 8-7), where oversized screens can contain thousands of lines and hundreds of colors. It is even possible to draw a three-dimensional picture of the manufacturing system with a detailed enough simulation to comprise an animation which might not be distinguishable from a camera view of the real system.

The detail available in a CAD system goes beyond functional needs of animation in the role to study the efficiency of control algorithms for hardware configuration. All that is necessary is to provide a means by which the situation can be displayed and the resulting action from a control algorithm can be observed. For functional purposes, the use of simple images in the microcomputer provide just as much detail and require less time and expertise to use than the elaborate CAD systems.

8.2.4 Detail of Simulated Activity

Another area where detail is of importance is in the amount of action which is displayed in the animation. The animation can only show activity which has been included as part of the simulation. For example, if the transportation system is simulated as a single resource requiring some predictable time to pick up and deposit a part, the animation, seen in Fig. 8-8, will only show the part move into the transport system and remain there until its movement time has elapsed. But if the transportation system is simulated as individual carts stopping and starting independently of each other, this activity detail can be shown in the animation (Fig. 8-9). If the detail is not visible in the animation, it is because it was not included in the simulation.

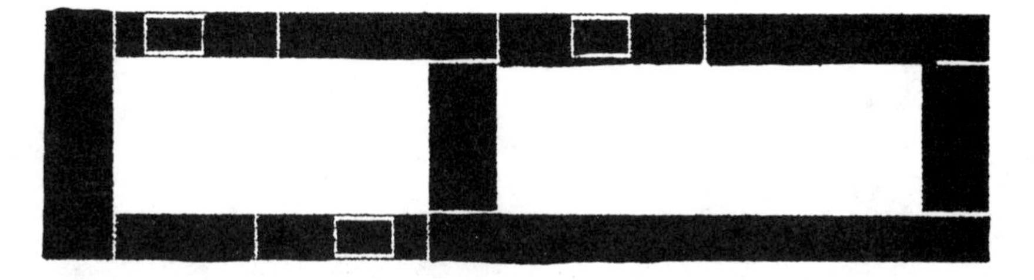

FIGURE 8-8 Simulated transportation system.

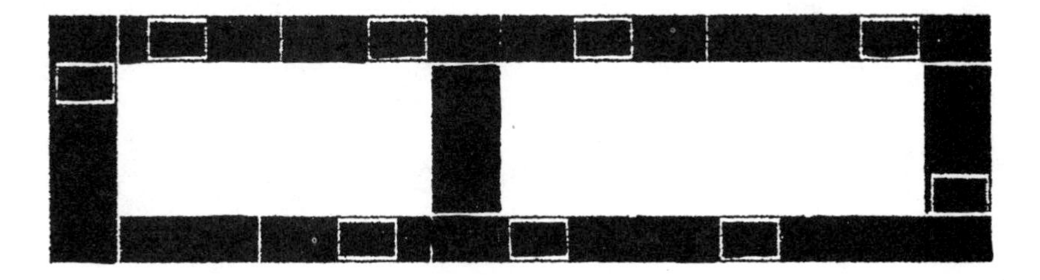

FIGURE 8-9 Detail of simulated transportation system.

8.3 ANIMATION'S ROLE IN DESIGN

Once the characteristics of animation are understood and it is possible to blink images across a computer screen, it is necessary to define what it can and cannot do. Its role in the design of a flexible manufacturing system must be understood because it does not replace the need for capacity planning nor simulation but adds value in some specific ways.

The specific feature of animation is that it permits analysis of minute-by-minute decisions. These decisions include which stations should be assigned to a part under the current situation, or which cart should be assigned under the current situation. Thus, animation provides the ability to analyze "situation decisions" (Fig. 8-10).

8.3.1 What Animation Does Not Do

These situation decisions will be described in the next section, but first, some discussion of what animation cannot do is needed. Animation cannot provide information for long time-frame decisions.

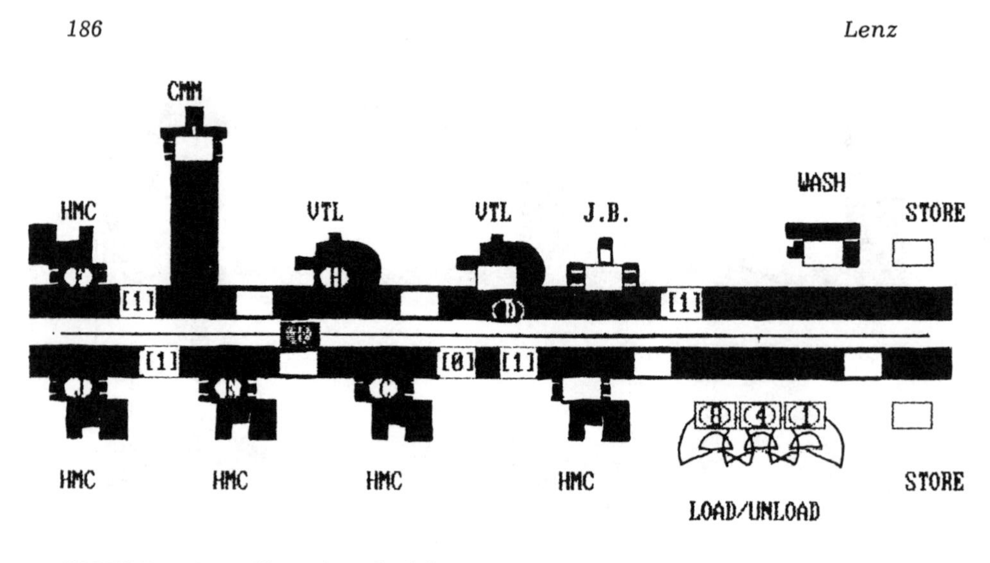

FIGURE 8-10 Situation decisions.

These decisions include utilization, production rate and other performance measures. These measures are collected over long periods of time, usually hours or shifts, and reported as a summary at the end of the collection period.

An animation can be viewed for an hour of simulated activity, but the viewer will not be able to give utilization statistics or even production rates. Of course, many animation programs provide these statistical results in conjunction with the animation. But these are no different than those statistics which were described in Chapter 7 as typical simulation output. This chapter describes the benefits of animation to flexible manufacturing design.

Therefore, animation provides useful information for situation type decisions but cannot replace the need for summarized performance statistics. It is best suited to reviewing the efficiency of control algorithms for a variety of situations. Computer simulation is excellent at constructing all kinds of situations, but performance statistics only summarized or average the effectiveness of a decision to given situations.

8.3.2 What Animation Does Do

Animation provides the ability to analyze control algorithms effectiveness on a situation basis. In some situations, the FMS performance might be acceptable in all areas except when one station group's utilization is below targeted levels. Integration effects prevent the station group from attaining its desired level of activity, but the solution might not need to be a shift in inventory or reduction of flow time. In this case, the integration effect might be due to an inefficiency in the control algorithms. But performance statistics cannot provide useful information because this effect must be analyzed on a situation basis or case-by-case analysis. This is where animation can be most effective in flexible manufacturing design. The control algorithm which can be evaluated on a situation basis includes station selection, traffic control, storage control and scheduling. Each of these will be described below.

8.4 MEASURING EFFECTIVENESS OF STATION SELECTION ALGORITHMS

Station selection is the process of assigning a part to a station. This decision is made when a part has just completed an operation and its next operation has been determined. At this given moment in time, the current situation of station status is used to determine which station gets the assignment. The optimal decision is to select the station where the part will be delayed the shortest amount of time. For this reason, many station selection rules use the current backlog of assignments to determine the priority of each station. With the use of an algorithm, summarized in Fig. 8-11, its effectiveness will be averaged for all situations and reported in terms of performance statistics. But this will not indicate whether the algorithm produced "optimal" solutions for all situations.

If a station's use is below targeted levels, it might be that the algorithm is not providing suitable decisions for some situations. In this case, animation of the system can provide a visualization of situations and resulting decisions. Various situations can be reviewed individually with immediate evaluation of results. There is no signal for when these situations might occur in the animation, so careful study is required for long periods of time. Also, the speed

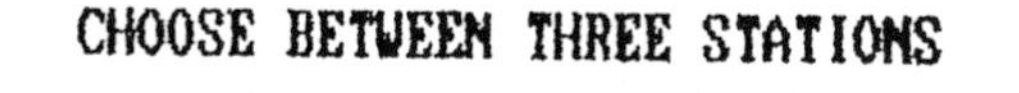

FIGURE 8-11 Alternative algorithms for station selection.

at which the animation runs must be controlled so that time is permitted to understand the cause and effect relationships which are creating situations.

When evaluating station selection algorithms, as parts complete an operation, the visual status of stations which are suitable for its next operation must be noted. From this visualization, the viewer should identify the best station to use. As the animation proceeds, the viewer can measure his decision against the control algorithms selection. If the two differ, a logical reason must be established and might provide information which can embellish the station selection control algorithm.

This method of viewing situations, anticipating a result and observing the action from the control algorithm can be applied to other decision areas. Another algorithm which needs evaluation for its effectiveness in various situations is traffic control.

8.5 MEASURING EFFECTIVENESS OF TRAFFIC CONTROL ALGORITHMS

Traffic control is the process of assigning vehicles and paths to meet the requests of parts. Animation can provide the ability to isolate cart and path selection into separate decisions, as shown in Fig. 8-12, or provide the merging of this into a single decision.

8.6 MEASURING EFFECTIVENESS OF STORAGE CONTROL ALGORITHMS

A third control area in which animation can provide an effective evaluation tool is in *storage control*, depicted in Fig. 8-13. Many system designs have provided ability for parts to be moved to dedicated storage facilities between operations. This permits free movement of parts but adds variability to their flow time. Performance statistics provide overall effectiveness of storage strategies, but it might only be needed for certain situations. Animation can be used to single out those situations where temporary storage of parts is necessary. This tailoring of algorithms for specific situations is important to maintain efficiency for given system configurations.

8.7 MEASURING EFFECTIVENESS OF SCHEDULING ALGORITHMS

A fourth decision area where animation can assist in measuring the effectiveness of algorithms is in *part scheduling* (Fig. 8-14). This

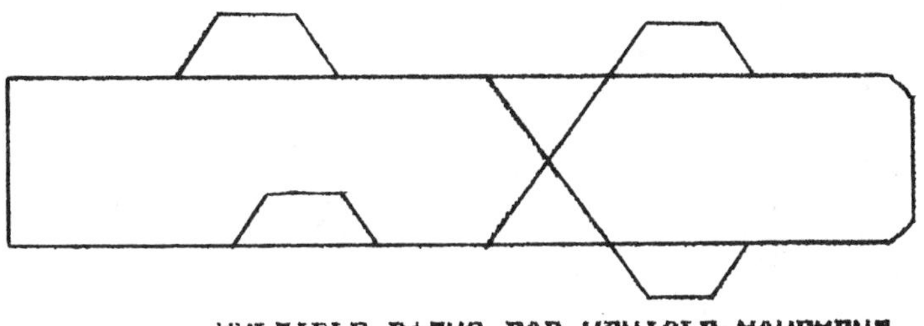

FIGURE 8-12 Alternative algorithms for traffic control.

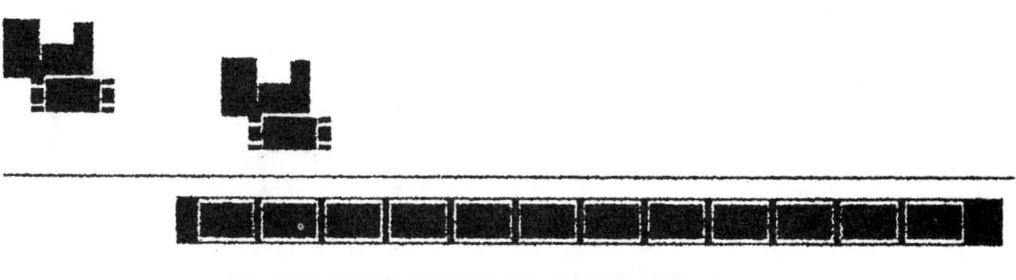

FIGURE 8-13 Alternative algorithms for storage control.

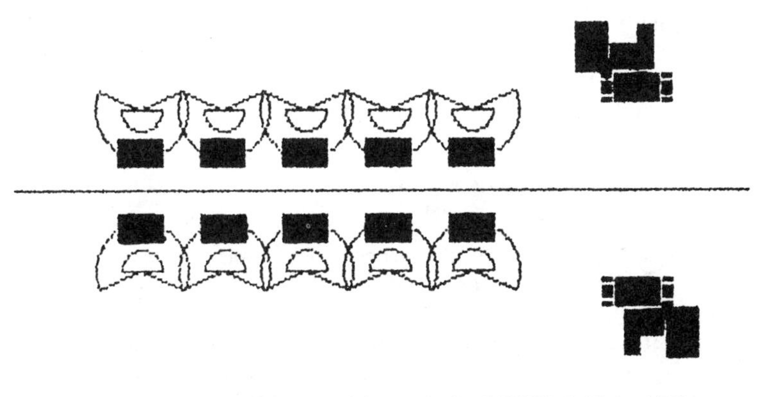

FIGURE 8-14 Alternative algorithms for part scheduling.

algorithm provides the sequence of parts which are introduced into the flexible manufacturing system. (In the pure sense, scheduling assigns specific times to a sequence. Time is impossible to control in an automated FMS, so most "scheduling" algorithms are only providing a sequence for which parts are to be introduced. But because the term "scheduling" has become a common one, it will be used to refer to establishing a sequence of parts for the FMS.)

The introduction of a part into an FMS occurs whenever a pallet, fixture, part and load station are available. The variability of this decision is then based upon the number and variety of parts waiting to be introduced, and the flexibility in the fixturing. In many designs, the effectiveness of a scheduling algorithm can be limited by the physical characteristics of the system.

When alternative parts can be introduced, the choice of which part can be substituted will be reviewed using animation. The summary statistics will provide long range effectiveness of an algorithm, but in order to identify weaknesses, specific situations need to be analyzed.

Animation can provide current statistics of all stations and visualization of all pallet positions can provide estimates of backlogs for each station. This information will permit the viewer to anticipate which part would be best to introduce next. This anticipation can be compared to what actually happens in the animation. If these two differ, some logical explanation is needed with the discussion resulting in a possible improvement to the algorithm used for introducing parts.

8.8 USER'S ROLE IN ANIMATION

One natural question to be raised at this point is why the viewer is not allowed to make the decision and inject it directly into the simulation. There are two reasons why this type of facility can be ineffective for solving control problems.

8.8.1 Animation Ineffectiveness

The first way that animation can become ineffective when allowing viewer injection of decisions is that the process can become nonexperimental. By definition, an experiment is any process which can be observed and replicated. The decisions of the viewer can be observed, but it might be impossible to replicate them should one small change need to be studied. Even if viewer decisions could be replicated, there is still a need to transfer them into an algorithm for automated decision making. This becomes the second problem with viewer-injected decisions.

In the actual operation of an FMS, control algorithms will be used to make decisions, unless, of course, the viewer is able to view the real FMS and inject decisions. Therefore, when the viewer is allowed to inject decisions, he must be able to transfer this process into an algorithm. Better yet, he should use an algorithm! But if he is to use an algorithm, it might as well be programmed and used as part of the simulation. In this way, the developer of the algorithm can test his algorithm for a variety of situations and observe the resulting decisions. In some cases, it does not have to be the algorithm developer, but anyone who understands how the FMS should operate can assist in its control evaluation using animation. In this role, the viewer is not able to actually program the algorithm (nor is it necessary that he do so), but rather he can communicate situations where improvements to the algorithm are needed. Here, animation provides a common communication tool from which all who are interested in control efficiency can identify specific situations and discuss desired decisions.

So far, animation has been treated as a tool for evaluating control algorithms in simulation generated situations. But its role is not limited to this one. Animation can provide usefulness for involvement in flexible manufacturing design and operation.

8.9 ADDITIONAL ROLES OF ANIMATION

Animation provides the ability to evaluate control algorithm effectiveness on a situation basis. This is the most vital and true engineering role. However, color graphic animation of simulated activity has many other roles in flexible manufacturing systems as well.

8.9.1 Presentation of Management

One of the most widely used roles of animation is to provide effective communications for management. A logical transition is needed from system description to performance measures. Often, an FMS is described in terms of parts, stations, process and layout, and other physical characteristics. Immediately following this information are the performance measures from the system. There is a big gap between individual components and integrated performance, and management needs a logical connection between the two. Animation can bridge the physical description with the performance measures.

As the animation runs, management can observe the integration, seen in Fig. 8-15, and the detail which was included in the evaluation. In most cases, they only need to observe a few minutes of an animation. Once they have seen the integration, they are interested in performance measures. In fact, long discussions about production

Additional Roles of Animation

Primary role — control algorithm effectiveness
* Visualize Component Integration
* Build Confidence in Simulation Results
* Validate Simulation Activity
* DeBug Real-Time Control Software

FIGURE 8-15

rates, individual station sensitivity to production and flexible paths of production usually follow the viewing of an animation.

Animation in itself, then, does not provide many of the answers to questions asked by management. Rather, it establishes a logical link between individual components and the integrated FMS. Once this link has been established, management is able to discuss system performance at length.

8.9.2 Confidence of Results

Another role of animation is tied to confidence in the results. Computer simulation is known for providing volumes of statistical measures, but the accuracy of these results has never been provided. In some cases, judgment and a gut feeling are more relied upon than simulation results in evaluating a flexible manufacturing system. This is usually due to a lack of confidence in simulation results.

Animation, with its ability to visualize the detail used in the simulation, provides a means where confidence can be gained. By observing an animation, the viewer can establish some "accuracy factor" as to what he sees. He will apply this confidence to the simulation results. Confidence in a result always needs an interval and animation can provide this interval.

Confidence must be established for the decision maker but must also be gained by the simulation developer. Because all simulation involves use of a mathematical model to represent physical objects, confidence must be gained in the model.

8.9.3 Model Validation

The process of gaining confidence in a simulation model is called model validation. Validation establishes a logical link between numbers inside of a computer and real situations. If the numbers do not accurately represent physical situations, a model is considered to be invalid and the results are meaningless.

In many cases, model validation is achieved through use of event traces and use of manual tracking systems. This process, however, was time-consuming and reserved for the simulation user himself. In this situation, other users of simulation results were dependent upon the expertise of the individual who produced the simulation. In this way, confidence in simulation results were tied to the confidence in the simulation expert.

8.9.4 Accuracy of the Simulation Study

Animation provides a role where all users of simulation results can judge the accuracy of the simulation model. Therefore, model validation is no longer a process done by one individual in a closed room. Rather, animation brings visibility so all users can participate in model validation.

8.9.5 Debugging of Control Algorithms

One final area chosen as a role of animation is in the debugging of real time control software. The primary problem with writing control software is that the physical hardware is needed to "close the loop." In other words, the software can issue decisions, but until some status comes back, additional decisions cannot be made. Simulation provides a means to close the loop by returning status information.

This allows the control software to be operated before actual installation in the FMS. Upon completion of this exercise, the software should operate for the situations which were tested. A better method of testing control software is to have a simulation return status but also create various situations. When a simulation can create situations, the control software can be studied in an environment much closer to the one it will eventually need to operate in.

Animation is used in this role to provide visualization of situations created by the simulation and decisions instructed from the control software. This method of debugging is identical to that used in the actual flexible manufacturing system. In the FMS, situations will arise and software developers will observe the software's reactions. If the reaction is not desirable, enhancements to the software will be necessary. Animation provides a debugging environment which closely resembles its eventual role in the FMS.

8.10 CONCLUSIONS

Animation can be an effective tool for establishing confidence in simulation results and evaluation of control algorithms. In both of these roles, animation adds value to flexible manufacturing design and evaluation. Sometimes, however, the animation must show the FMS as having no lost efficiencies due to integration. In these cases, the animation is no longer a tool for FMS evaluation but becomes the problem itself. In making the animation look good, reality is often traded for efficiency. When this happens, animation can become one of the most expensive video games ever purchased. Color graphic animation has an important role in flexible manufacturing and care must be taken to ensure that its use remains at a professional level.

9

Benefits of Flexible Manufacturing Systems

Why choose flexibility over other manufacturing techniques? What real benefits will flexibility provide? These are two common questions and their responses always focus on strategic issues. These issues include product life cycle, direct labor input and market characteristics. It is true that flexible manufacturing will affect these strategies, but the alternative to flexibility will also influence them. The real issue is to quantify the benefits and compare these measures of each alternative strategy.

Rather than repeat the traditional list of opportunities which flexibility offers, a method is described from which quantified benefits can be derived. This provides a framework from which alternative manufacturing techniques can be compared. The following sections contain a description of these economics of manufacturing.

9.1 ECONOMICS OF MANUFACTURING

The economics of manufacturing are no different than any other business investment mode. The rationale for any investment is to obtain an economical rate of return on an investment. Economists might refer to this characteristic as the "bang for the buck" measure or more commonly referred to as the Return On Investment (ROI), seen in Fig. 9-1. In terms of manufacturing, the investment is the cost of acquiring and operating the production facility. This cost is comprised of equipment acquisition, operating, labor, material and management cost. The return of this investment is realized as the "additional value" and quantity of products which are produced in the facility. The additional value is the difference between

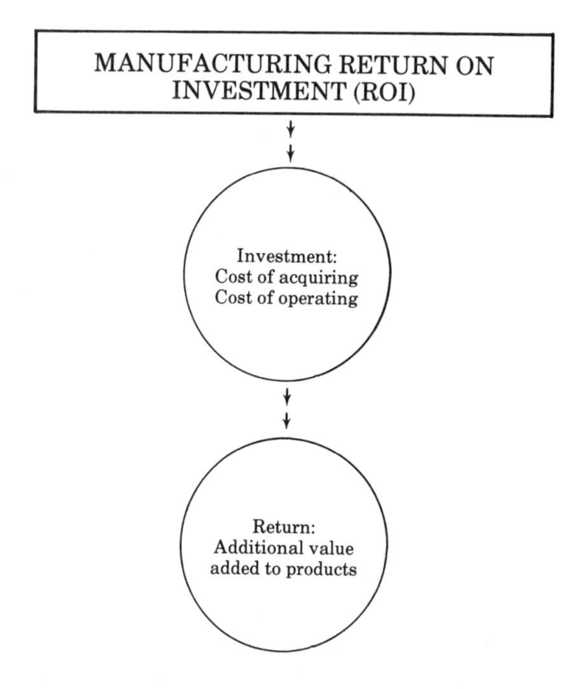

FIGURE 9-1 Manufacturing ROI.

the selling price of the product and the cost of acquiring raw materials. The quantity of products produced is the productivity of the manufacturing facility. To measure the productivity requires computing the real or net production capacity of the facility. The following sections contain a description of the factors which affect investment and productivity of a manufacturing facility.

9.1.1 Factors Which Influence Manufacturing Investment

The investment in a manufacturing facility represents the capital which is needed to produce the desired product(s). This investment will be influenced by several factors, ranging from the variety of products which are to be manufactured to the instinctive nature of the manufacturing decision makers. These factors are listed below and summarized in Fig. 9-2.

One factor which influences the investment in the manufacturing facility is the volume of products which are to be produced. This volume has been categorized as either high, mid or low. High volume is usually used to describe products which are used in the

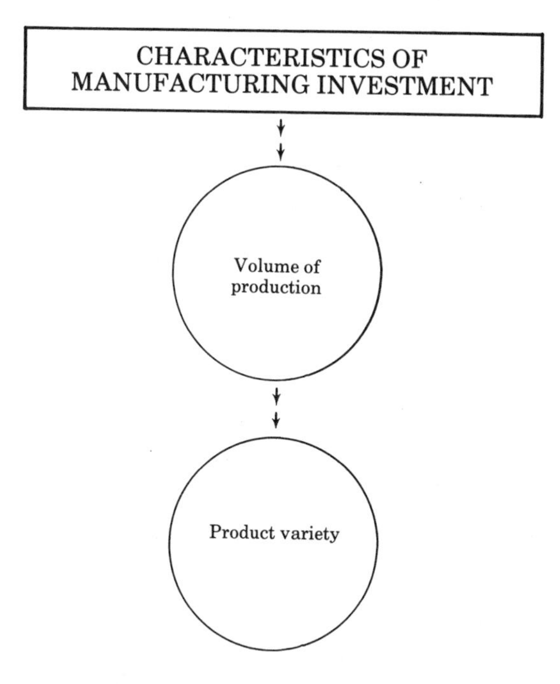

FIGURE 9-2 Manufacturing investment.

automotive industry. For example, the automotive industry produces more than 100 automobiles each hour. Mid volume manufacturing is found in the construction or aerospace industry. In these industries, production can range from 10 to 100 units per hour. Low volume manufacturing is found in the manufacture of highly technical industries such as medical equipment. In these industries, production is less than 10 parts per hour.

Volume by itself does not have a uniform relation to investment. For example, in a high volume production facility, the amount of equipment needed might be less than that in a low volume production facility. This will depend upon other factors such as product variety and the process itself.

The second factor which influences the investment is the product variety. Product variety is used to describe the differences between products which are to be produced in the same facility. This variety can range from small variations of a "core" product (these might be referred to as family products) to products which have no shared resources in the facility.

As with volume, product variety by itself has no uniform relationship to investment. For example, a facility which manufactures

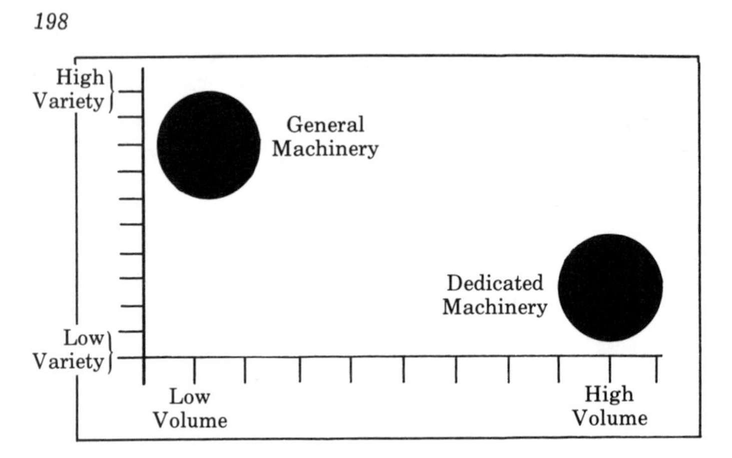

FIGURE 9-3 Production volume vs. variety.

a single product could require a greater investment than one where hundreds of different products are produced. However, the combination of variety with volume has provided some general statements on their combined relationship to investment.

The relationship between variety–volume factors and investment is established by use of the type of equipment which is commonly found economical. Figure 9-3 contains a common method for showing the relationship between the combination of product variety and volume with the nature of the equipment used.

For an equivalent amount of production capacity, the more dedicated equipment will require a lower investment than general purpose equipment. It is in this distinction where the combination of volume and variety, through the type of equipment which is commonly used, influences the investment of the production facility.

The investment is sensitive to the type of equipment used in the facility. Besides the volume and variety factors, another factor which has a direct effect upon the equipment used is the process selected for the manufacturing of products. The process includes such characteristics as the number of subassemblies used in the completion of the part to the use of specialized tooling instead of a more general tooling to carry out production.

The choice of volume or variety is usually influenced by market opportunities, whereas the choice of the process is a factor where many alternatives can be explored. For this reason, the choice of the process has a dominant impact upon the investment within the production facility. Because of this important characteristic, the proposed evaluation procedure is based upon the need for the evaluation of alternative processes. But before this evaluation procedure is described, one must examine factors which influence productivity.

9.1.2 Factors Which Influence Manufacturing Productivity

When deciding which factors to describe for investment and productivity, the degree of automation clearly affects both. The higher the degree of automation, the greater the investment because of the computerized control systems which are needed to provide the feedback into the equipment. This feedback is a replacement of some direct labor, but from experience the direct labor saved is replaced with indirect labor for support. Because of this conversion of direct to indirect and the need for a control system, automation does require a higher investment.

The economics of manufacturing are not based upon investment alone, but on the return received from the investment (Fig. 9-4). In these terms, automation will be economical when improvement in

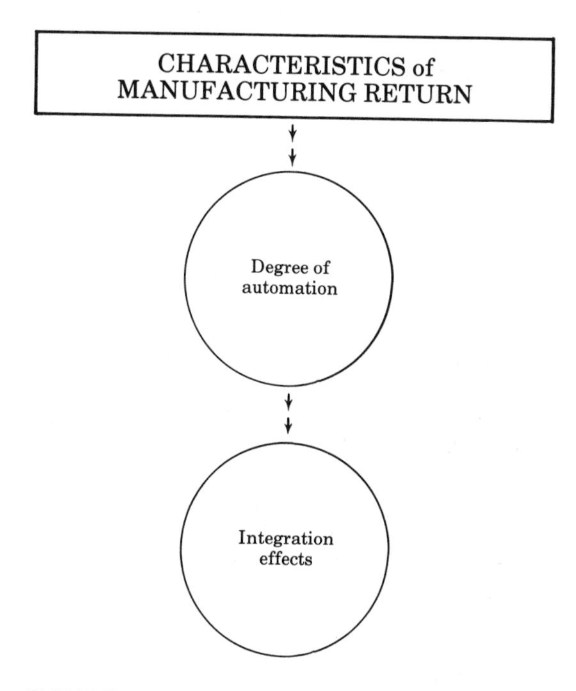

FIGURE 9-4

productivity outweighs higher investment cost. The study of this relationship is the focus of the evaluation procedure. Factors which influence the productivity of a manufacturing facility are derived from the Manufacturing Integration Model (MIM).

Productivity or net production are defined by MIM as a function of gross production capacity, station availability and integration effects. Gross production capacity is determined from the process requirements and the specific equipment which is selected. The unavailability of the equipment subtracts production from the gross capacity level. This downward adjustment is well understood and techniques are available for anticipating the extent this will have upon productivity (real production). This technique for establishing the relationship between station availability and productivity is through the use of efficiency factors. These factors are based upon experience and have proven to be a reasonably accurate technique.

Where the relationship to productivity is not easily identified is with the integration effects factor. These integration effects are the carry-over effects from individual station interruptions. According to MIM, the degree to which these integration effects have an effect upon productivity is determined from inventory level, balanced loadings of equipment and flexibility. A procedure for measuring the integration effects adjustment to productivity which involved computer simulation was described in Chapter 3 and Chapter 7.

Because of these factors' influence upon productivity, the evaluation of economic manufacturing requires a procedure which can produce comparable returns on investments for manufacturing alternatives.

9.2 ECONOMIC EVALUATION PROCEDURE

The first step in the evaluation of economical manufacturing is to define and specify a process for manufacturing the product. The choice of potential processes is based upon the experience and intuition of manufacturing management. Once the process is defined, the second step is to identify suitable equipment to carry out the desired process.

This second step is usually performed with the assistance of equipment vendors and other experts. This equipment list includes the definition of each work station, a layout (relative placement of the work stations), a material handling system, a material storage facility and control systems. This control system will be described either as computerized or manual procedures. These proposals must then be evaluated for net production and productivity.

ECONOMIC EVALUATION PROCESS

Step 1: Define and specify a process (time study)
Step 2: Obtain equipment list
Step 3: Determined productivity
Setp 4: Define costs
- Acquisition cost
- Labor cost
- Operation cost
- Indirect/management cost
- Work in-process inventory cost

Step 5: Define return
- Identify net reduction using MIM
- Added value to product

Step 6: Compute return on investment ratio

FIGURE 9-5

The third step is to identify the productivity of the proposed manufacturing facility. This evaluation is based upon MIM which states that some combination of work in-process, balance loads and flexibility will yield minimum integration effects resulting from interruptions in station availability. This step will also establish how sophisticated the control system must be to manage the work in-process, balance (scheduling) and flexible routing.

When the variables which yield the highest productivity have been identified, the next step is to compute the investment cost of the proposed manufacturing facility. The investment will include the cost of acquiring the physical components of the facility, the inventory cost associated with the work in-process level, the cost of the control system, estimated operating costs, installation costs and management costs. The equipment acquisition cost is the total investment for machinery, transportation and storage facility. The inventory cost is found from the value of the work in-process material and some adjustment for interest rates. The cost of the control system includes computer and software acquisition if a computerized control is used or the cost of manual record keeping for noncomputerized control. The operating cost will include the direct labor cost, maintenance labor input, spare parts inventory and consumables, such as tooling or coolant. The installation cost includes the actual installation plus costs for interruptions in service and elapsed time before the facility reaches its level of productivity. The last major investment component is the cost of management. This cost includes indirect labor for engineering support and general management. The total of all of these costs represents the investment in the manufacturing facility. But not all of these costs are fixed costs. Some will be based upon a given time period and will continue throughout the life of the manufacturing facility. The characteristics of some of the costs must be taken into consideration and is usually done through totally all variable costs for some return on investment window. For example, the return on investment window might be five years. In this case, the total cost for acquiring and operating the facility for five years would be used as the investment.

The final step in the economic evaluation of manufacturing is to identify the production which can be realistically obtained in the return on investment window. With this value, the return on investment can be computed by identifying the net revenue which can be expected from the production and the total investment. The net revenue is the difference between the acquisition cost of the raw material and the revenue from selling the products, multiplied by the real production. The return on investment is the net revenue divided by the total investment.

This bottom line result should be computed for a variety of processes as well as equipment alternatives. The process should range from high degrees of automation to others with manual operation. Equipment selection should range from dedicated equipment to highly flexible equipment. But these choices will depend upon the insight of the manufacturing management.

The insight of the manufacturing management will be based upon their experience and understanding of the relative impact that alternative processes, automation, inventory levels and flexible equipment have upon investment and productivity. The following section contains a description of the relationship between flexibility and these measurements.

9.3 BENEFITS OF FLEXIBILITY

The benefits of flexibility are presented in terms of common objectives which are derived from the manufacturing facility. Each objective is listed with the specific relationship flexibility can play in achieving this objective. The intent of this description of flexibility in terms of goals is to provide insight into the complex effects flexibility has upon investment and productivity.

9.3.1 Objective: Maximize Production

The objective of maximum production can be interpreted by use of MIM. To maximize production for a given amount of gross capacity, the amount of station unavailability must be minimized as well as the carry-over effects from these interruptions in service. Minimizing station unavailability will require short repair times when a station fails and preventive measures to reduce the frequency of these interruptions. Short repair times can be accomplished with extensive training of maintenance personnel and by maintaining an adequate supply of spare parts. To obtain shorter repair time requires a large investment in operating expenses.

The second method for increasing station availability is to prevent the station from failing. This method is referred to as preventative maintenance where routine checks and diagnosis are performed on the stations. Many strategies for preventive maintenance use third shift or off-hours as a time to remove a station from service and perform diagnostic testing. Any method of prevention maintenance will require a greater investment.

The cost of keeping stations from failing and ensuring short repair times has diminishing economies. That is, the cost of reducing a repair time by 10% will cost more as the time gets shorter. Due to this characteristic, some interruptions in service will still occur. In many instances, the lost production which is directly attributed to the interruption is less than the lost production due to the carry-over or integration effects which result from the interruption (Fig. 9-6).

According to MIM, the amount of production which will be lost to integration effects will be determined from the current work in-process level, degree of balanced loads across the work station and flexibility. Therefore, the greater the degree of flexibility in the

BENEFITS OF FLEXIBILITY:
Maximize Production

Flexibility reduces lost capacity due to carry-over effects from station failure.

FIGURE 9-6

production facility, the lower the integration effects and the higher the net production. As parts travel to more places for operations, less traffic congestion will result and station utilization will be higher.

If the objective is to maximize station utilization or net production, flexibility can lower carry-over effects when stations interrupt their service. Flexibility can also lower the cost of maintaining spare part inventories due to the fact that similar equipment can share components. Flexibility can also benefit preventative maintenance by permitting it to be performed during regular production hours. This will reduce the cost of off-hour labor. In summary, the higher the degree of flexibility of the work station, the lower the potential cost of production capacity due to station unavailability.

9.3.2 Objective: Just-In-Time Manufacturing

Just-In-Time (JIT) manufacturing has been promoted as an important technique for future manufacturing trends. The manufacturers in the Far East have proven that these techniques can be implemented and provide substantial economic benefits to manufacturing. This has prompted many Western manufacturing organizations to consider JIT as part of their manufacturing objective.

Just-In-Time provides a technique where inventory can be exchanged for sophisticated operational control procedures. In terms of MIM, JIT techniques with use of control procedures must manage the increase in integration effects which results from the lower inventory level. However, control strategies can only provide an "after the fact" solution. In this sense, control is a response to something which has already happened. For this reason, control will have limited impact in reducing integration effects.

The limitation of control is explained through the use of an example. A manufacturing company decided in 1983 to implement some JIT techniques with the primary objective of lowering work in-process levels. After three months, the work in-process level was reduced by 50% from what it was prior to JIT implementation. As a side effect, however, the production level of the facility dropped by 20% as well. This drop would have been acceptable but net production was already at market demand level. Therefore, gross capacity was increased by acquiring additional labor for a third shift operation.

From this example, JIT did reduce the work in-process level, but it was not able to control the increase in integration effects which resulted from lower inventories. The integration effects increased, which lowered the net production of the facility. Rather than reduce integration effects, this company elected to increase

BENEFITS OF FLEXIBILITY:
Just-In-Time Manufacturing

Flexibility is used in exchange for inventory to reduce integration effects.

FIGURE 9-7

the gross requirements by adding a third shift of labor to increase net production. This is a common reaction to many JIT installations.

When integration effects are managed by use of flexibility, the inventory does not have a dominant impact upon integration effects. Therefore, techniques which reduce inventory levels will not reduce net production indirectly in flexible production facilities. In other terms, when inventory is reduced, flexibility must be increased to maintain the same level of net production (Fig. 9-7). This point can be further explained by use of the importance of flow time to integration effects. For JIT to maintain the same net production levels with lower inventories, the flow time for parts must decrease. This relationship between inventory, flow time and production is described through use of the WIPAC Curve. In order for JIT to maintain net production, flow time must be fast (parts must have an increased velocity) and this velocity must be stable. The fast flow time will result from lower inventory by having fewer parts to compete for limited resources. The control system is used to control the variability of the flow time. But the control strategy is limited in its ability to control variations of flow time. However, small variability to flow time is a characteristic of flexibility.

With an objective of reducing work in-process by use of JIT techniques, net production will not decrease if the variability of flow times does not increase. Control strategies can attempt to manage this variability but their impact is limited. Flexibility offers an effective means to maintain low variability to flow time. As a result, JIT techniques will reduce net production for those production facilities which use flexibility to manage integration effects.

9.3.3 Objective: Make Every Product Every Day

Manufacturing organizations which produce a variety of simulation products with use of the same equipment are looking for techniques which reduce their dependence upon batch production. One of

BENEFITS OF FLEXIBILITY:
Make Every Product Every Day

Flexibility provides minimum set-up delays during reallocation of resources.

FIGURE 9-8

their objectives is to reduce setups so that every product is produced every day. In this sense, they are moving away from a batch production environment to one that resembles continuous production. But it can never be purely continuous because discrete parts are required in fixed amounts.

In order to maintain net production and make every product every day, the production facility needs to allocate its resources to match the functional needs of the products (Fig. 9-8). This allocation will require work stations to be able to perform several different manufacturing functions. However, each time that the work station changes its functions, it incurs a setup delay. For this reason, economic lot sizes are based upon the trade-off of inventory cost and setup cost.

Flexibility provides a means by which to reduce these setup costs in two ways. First, flexibility offers work stations characteristics which, by their nature, have low setup times. Secondly, the stations which are flexible can usually perform a variety of manufacturing functions. This permits the ability to adjust capacity to match production needs. For example, one mix of products might require five hours of operation A and ten hours of operation B. But this mix changes its load or demand of operation A, and B will change as well to perhaps nine hours and two hours. If some of the operation B stations could be set up to perform operation A, efficient use of stations could be maintained. To make every product every day requires that the capacity of the production facility be adjusted to meet the changing needs due to product mix. Flexibility offers this characteristic to the production facility. Flexibility is also compatible with a high degree of automation which can eliminate setup delays. This setup elimination is accomplished by computer tracking of parts and an automated material handling system. These characteristics have a high cost and the economics of their application must be studied with the procedure described above. But, despite the degree of automation, flexibility provides for characteristics in the manufacturing facility which are essential for researching the objective of making every product every day.

9.3.4 Objective: Maintain Less Than Five Percent Rework or Scrap

An objective to minimize the number of parts which require rework or which must be scrapped implies that the manufacturing process must be kept within some tolerances. The frequency of making a bad part will be determined from the process and nature of the operation of the equipment. This objective is not intended to prevent the occurrence of making a bad part but rather to make only a few bad parts when an operation is out of tolerance.

The method for reducing the number of parts which require rework or need to be scrapped requires that inspection of operations be moved from a postposition to an in-process position. That is, the parts are inspected as they are produced and not intermittently at the end of a series of operations. This in-process inspection is accomplished by using a procedure where inspection is performed immediately following an operation and feedback is provided to maintain the process within control limits.

The repositioning of inspection throughout a process requires that either the work station performs to its own inspection or some inspection operations follow at a different station. This need for tracking, collecting of process control data and the important use of feedback is the reason why computerization of the control is often used.

It is with this need for computerization that flexible manufacturing beneficially reduces rework (Fig. 9-9). Flexible manufacturing is very compatible with computerization and, in fact, most FMS use computerized control systems. This control system provides the

BENEFITS OF FLEXIBILITY:
Maintain Less Than 5% Rework

Flexibility lends itself to computerized tracking of work flow which is helpful for positioning inspection throughout a process.

FIGURE 9-9 Benefits of flexibility: maintain less than five percent rework or scrap.

data collection and distribution network which can serve the SPC technique. If the control system is not computerized, operators must collect data, and a formal information system must be established to feed back this information into the process. Operators often do not see the need for accurate, timely data collection and often feel that carrying out an operation is more important than taking time for accurate, timely reporting of data. For this reason, the success of SPC when using manual techniques requires the design and strict management of a formal information system. Such a system is already in place with the FMS computer control system.

9.4 CONCLUSION

The benefits of flexibility in the production process can be summarized as an alternative to traditional techniques for managing integration effects. The job shop is able to achieve low integration effects through high inventory levels. The transfer time is able to achieve low integration effects through a balancing of operations. However, competition has proven that there are substantial benefits for low inventory or unbalanced operations.

Today, control methods via scheduling only provide a means of reaching low inventories or unbalanced loading. However, these techniques fail in maintaining integration effects. Flexibility offers a real alternative to inventory and balancing which maintains integration effects.

Flexibility will not eliminate integration effects like inventory will in the job shop or balancing will in a transfer line. Therefore, net production will be lower in flexible systems than with comparative production facilities with high inventories or balance operations. To measure the benefits of flexibility requires a means to quantify the cost due to lower net production against the returns of low inventory and unbalanced loadings.

What mixture of inventory, balance loadings and flexibility provides the production facility with its most economic conditions? In other words, how much integration effects can exist before it is worth the cost of raising inventory or balancing operations? Flexibility does not come without its costs. However, it must be considered as an alternative technology which is compatible with the trends toward low inventory or unbalanced operations.

10

Cost of Flexible Manufacturing

10.1 INTRODUCTION

The cost of flexible manufacturing is the investment needed to obtain flexibility in a manufacturing process. The investment has been generally described as part of the economic analysis presented in Chapter 9. In this chapter, the investment will be described in more detail and will focus at specific characteristics rather than economic terms.

From this investment or cost of flexibility must come the benefits. These benefits of flexible manufacturing include a capability to respond to changes in the production environment. These changes or business diversification, to operational changes, such as a change in work in-process levels or in-process quality review. Flexibility provides its greatest ability in providing potential for adapting to most changes which might occur in the production environment.

Establishing these benefits from a cost or investment involves some risk as well. It is this risk of flexible manufacturing which complicates the economic cost/benefits analysis. The risk in flexible manufacturing is the chance or probability that the system will not achieve the planned level of productivity, or that the cost will be greater than expected (Fig. 10-1). When either of these change, the return on investment declines and thus the choice of flexibility becomes less desirable when compared to other alternatives.

The fact that the productivity or cost might change is not a primary concern. What is most important is the likelihood of change, and the degree to which they change. Flexible manufacturing comes close to the top in both categories; that is, carries a high likelihood

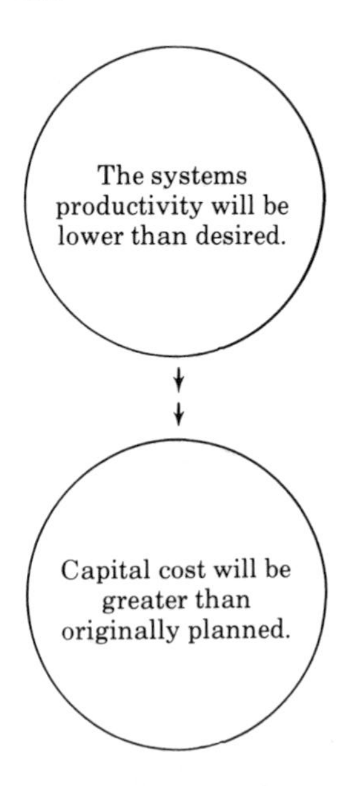

FIGURE 10-1 Risk of flexible manufacturing.

that the actual manufacturing system will not achieve the productivity which was called for in the original design. It is also likely that it will cost more than was originally estimated.

One means to reduce the investment in flexibility is to reduce the risk. This requires the reduction of the chance that the system will not reach desired production levels or cost overruns. This is the subject of the following section.

10.2 RISKS OF FLEXIBLE MANUFACTURING

The likelihood that an actual flexible manufacturing system will have a different production rate and cost from the original plan is quite high. This various return on investment components makes FMS risky. The way to reduce the risk is to first identify those characteristics which make up the risk.

These components can be categorized into three areas, illustrated in Fig. 10-2. First is the set of characteristics which provide the difference between planned production capacity and actual production capacity. The second set of characteristics are those which influence the cost of installation. The third set are those which

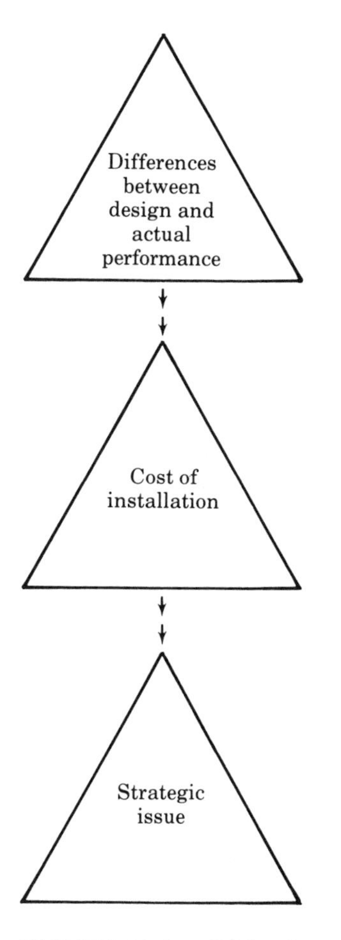

FIGURE 10-2 Risk characteristics.

deal with strategic issues. Each of these three sets is described in the following sections.

10.2.1 Risks in Design of Flexible Manufacturing Systems

The main outcomes of the design procedure for a flexible manufacturing system are the numbers and configuration of components needed to meet a desired production rate. The risk of this process is that the actual production from the number and configuration of components

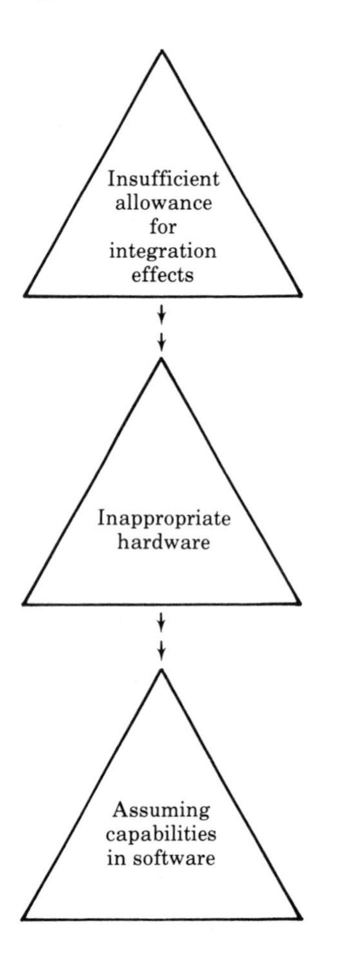

FIGURE 10-3 Risks in design of FMS.

will be less than the planned level. There are three primary reasons why this might occur, as summarized in Fig. 10-3.

First, the design process might not have sufficiently allowed for the integration effects of the FMS. These effects were defined as part of the Manufacturing Integration Model (MIM) in Chapter 2. In short, they are the carry-over effects due to low inventory levels, balanced loads and flexibility which are present in flexible manufacturing.

The design of a flexible manufacturing system involves mathematical models for capacity planning. From this starting point,

Do's and Don'ts

DO:
Use it to quantify integration effects
Use it to identify bottlenecks in the system
Use it to study control alternatives
Use it to prove a design

DON'T:
Use it to create unrealistic solutions to problems
Use it to force a concept to work
Use it to support mathematical calculations

FIGURE 10-4 Simulation's role in design.

computer simulation is used to replicate the potential operation. This simulation cannot be used for the purpose of verifying that the mathematical calculations were correct, but it can identify the differences. These differences are the integration effects and are real adjustments to the production capacity.

Instead of attempting to get these "abnormalities" out of the simulation through such procedures as intermediate buffering of parts, raising work in-process levels or achieving better balance loads, the design evaluation must be able to explain why these effects exist. Every FMS will have integration effects and it is the responsibility of the design to identify, quantify and explain these effects (Fig. 10-4). The role of computer simulation in FMS is proof of design. Therefore, most applications of computer simulation are to prove that the concept will work. Effective use of computer simulation for design is obtained when the simulation is used to identify integration effects and the design then deals with the effects. Adjusting a computer simulation to confirm the concept is one way by which the risk is increased in flexible manufacturing.

Other ways in which the risk in flexible manufacturing is increased occur when the integration effects are ignored. In actuality, they are not really ignored, but are assumed to be solved at the building time. There are two general approached to ignoring integration effects. These are the "kill it with hardware" solution and the "software can do it all" solution (Fig. 10-5).

The "kill it with hardware" solution is one where a weakness is recognized in the design and rather than change the concept, the

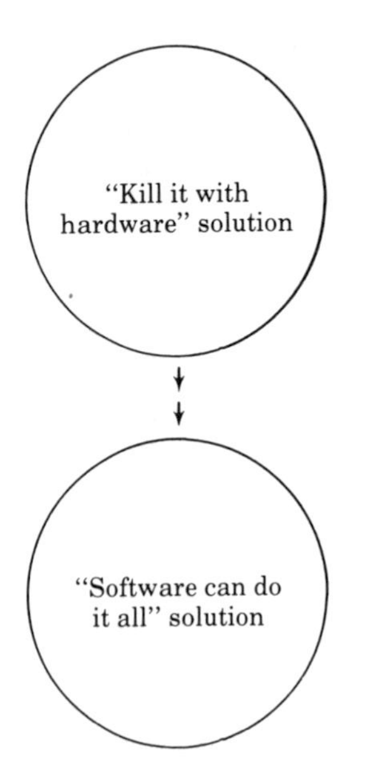

FIGURE 10-5 Methods to ignore integration effects.

hardware is changed. These hardware changes range from doubling the size of a component to adding functions which are completely different from its initial purpose. An example of this is in the chip removal from parts. Rather than increase an operation duration or change to different tooling (which might prevent the chips from becoming a problem), an elaborate vacuum system is proposed. The reason for choosing the hardware alternative is that integration effects will appear when the operation time increases. So rather than accept the fact that some integration effects will remain in the system, hardware is set in its place. The risks increase when this hardware is not a proven application and its interruptions in service create integration effects.

Often with the "kill it with hardware" solution, the design decision is to eliminate integration effects by the addition of hardware. However, the hardware often becomes specialized, and its interruptions in service create greater integration effects than those which would have resulted if they were left alone. The attempt to eliminate integration effects through unproven hardware application increases the risks in flexible manufacturing.

The second method of ignoring a flexible system's integration effects is the "software can do it all" solution. An example of this is most commonly found in the material handling system. The flexible manufacturing system is designed with complex flow patterns of a variety of parts. Rather than attempting to simplify flow patterns, it is assumed that the traffic control modules must be able to handle these patterns. The software is then left to solve an extremely complex routing problem, but as is the case with most FMS control systems, it only has information which is 50% accurate and timely. The software does not "optimize" the solution, so integration effects occur due to poor software. These losses in productivity due to the software are usually greater than the integration effects which would have occurred from attempting to simplify the flow within the flexible manufacturing system. The control software for FMS is the scapegoat for reductions in productivity. However, the software can only optimize a decision with perfect information. In addition, the software is required to eliminate integration effects and, if it does not, it gets the full blame for poor performance.

Control software will never eliminate integration effects because of the need for perfect information. Whenever the "software can do it all" solution is used to handle integration effects, the risk of flexible manufacturing increases.

These three causes for increasing the risk in flexible manufacturing can be avoided through the recognition that integration effects will exist within any flexible manufacturing system. The responsibility of the design is to minimize these effects through simple and proven solutions. However, whenever the design forecasts a greater than 90% efficiency and relies on software or unproven hardware to attain this, the risk of flexible manufacturing systems is high. The risk is that the real production rate will most likely be lower than that estimated in the design.

Not only are there risks in the production side of the return on investment analysis, but flexible manufacturing can have considerable risks on the investment side as well. These risks are described in the following section.

10.2.2 Risks in Installation of FMS

The risk associated with the installation of an FMS relates to the complexity and size of the projects being undertaken. It is common for an FMS installation to require twice as much time as originally estimated. The fact that there is such a high likelihood that the flexible manufacturing system's installation will be longer than planned can be explained through many different characteristics and is illustrated in Fig. 10-6.

One reason why the installation takes more time than planned is because of the high degree of integration found in flexible manufacturing. Each component within the FMS is the latest in manufacturing

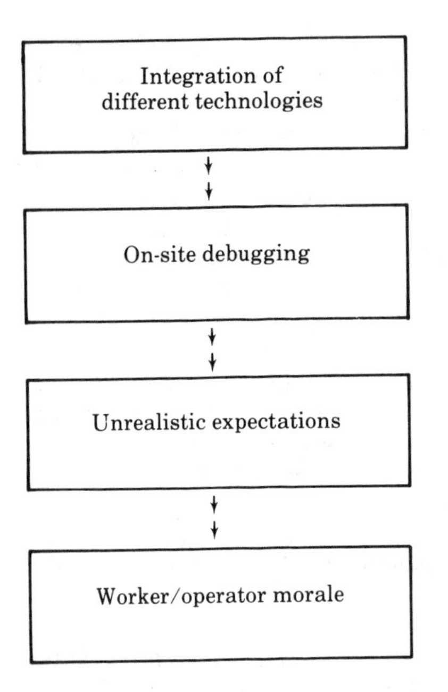

FIGURE 10-6 Risks in installation of FMS.

technology. With this high level of technology in each component, only people with specialized skills can provide effective assistance in debugging. These skills range from computer programming, programmable controllers, CNC executive programs, hydraulics, material handling mechanisms, storage—retrieval systems and other mechanical capabilities. The problem does not lie in the existence of these skilled people but in their availability at the appropriate moment in the FMS installation.

Each of the components can be individually debugged to some extent by the skilled people. The FMS installation is primarily comprised of debugging the integration of these components. This debugging of the integration involves a subset of these skilled people from time to time and it is never known who and when specific people will be needed.

Because of this, the FMS installation usually starts with all people at the installation site. But when problems are being debugged in the hardware, many skilled people are not needed. One by one, skilled workers are pulled from the installation and are placed in a "will call" position. When a need arises, they will be reassigned. This type of scheduling is usually based upon a one week window. This creates many delays and lost time waiting for the right person to solve the next problem. For example, a hardware integration problem might exist between the material handling system and pallet stand. While this problem is being resolved, people skilled in software are reassigned to other projects. When the hardware interface is resolved, the next step is to attempt to carry out the action under computer direction. This is tried and the first thing is that the software does not work properly. Now, however, the people skilled in hardware must wait for the software person to come back to the installation project.

The loss of a few days in a week based scheduling system can increase the installation time dramatically. As the installation is dragged out, the availability of the right skilled person at the right time usually becomes more of a problem. This in itself is a problem, but falling behind schedule usually results in a side effect, the shortage of funding.

Because of the size and cost of most FMS installations, only portions can be set up at off-site locations for testing. This leaves most of the integration and debugging to be done at the FMS site. Here, potential operators and other workers have access to observe the progress of the FMS. They will establish a perception of the complexity and capability of this new technology, but this perception will be formulated during the "trial and error" process of debugging. In most FMS installations, the operators are likely to arrive at work with mixed feelings which are based upon their observation of the installation.

This potential for low morale of operators increases the risk of flexible manufacturing. There are some procedures for reducing this risk; these are described in Chapter 12, which deals exclusively with FMS installations. The focus of this chapter is the added risks which are present in flexible manufacturing when compared to other manufacturing technologies. They have been presented in categories of design and installation. One third category which has obvious impact upon flexible manufacturing risks is the degree of automation.

Flexible manufacturing usually involves a high degree of computerized manufacturing. To obtain flexibility in a manufacturing process, the set up for different operations must be kept to a minimum duration. One method for reducing set up is through fixturized toolings, but this still requires tracking of the work flow. The tracking is most easily obtained through computer systems and with an automated material handling system.

From this evaluation of needs, it is clear to see that flexible manufacturing has become synonymous with computerized manufacturing. But computers are only one means to obtain flexibility (other methods were described as part of the benefits of flexible manufacturing in Chapter 9). Flexible manufacturing provides benefits to the characteristics of the production environment and computerization provides a means by which lower labor requirements realize these benefits. The degree to which an FMS is automated will increase its risk. However, this risk is not because of flexibility itself, but rather from the degree of computerization which was selected to implement it.

The degree of automation has an impact upon the operators as well. Most FMS have operators who have previous experience with operating complicated machinery. But in these previous jobs, they had responsibility to provide "hands-on" operation. However, when these workers become operators in a highly computerized FMS, they lose interest because their jobs change to loading or unloading parts, setting up tools and providing janitorial services. The enrichment of the job is given to the computer and operators are left with service tasks. In this manner, the degree of automation increases the risk of flexible manufacturing.

10.3 COST OF AUTOMATION WITHIN THE FMS

This discussion of flexibility and its distinct relation to automation is important when describing the risks of flexible manufacturing. It is also important to recognize that not all risks are due to flexibility, but some can be attributed to the method which is chosen to implement it. Not only is the distinction important in a discussion of use, but it is also important when describing the costs of flexible manufacturing. For this reason, the costs which are associated with automation are described separately from the costs which can be attributed to flexibility.

With any type of automation comes the use of computers. These computers provide the framework for the information system which directs action and monitors and feeds back the activities of the machinery. The cost of the machinery is approximately the same regardless of whether it is operated by a person or a computer. For this reason, when the cost of automation is discussed, it usually entails only the cost of computers and their integration. In an FMS, the most significant expense is that of the computer integration.

10.3.1 Cost of Computer Integration

Flexible manufacturing systems involve a wide variety of components, each with their own type of computer control. Many of these components are installed as islands of automation, each with a computer control capable of monitoring and directing the actions of the one component. These computer controls might include a programmable

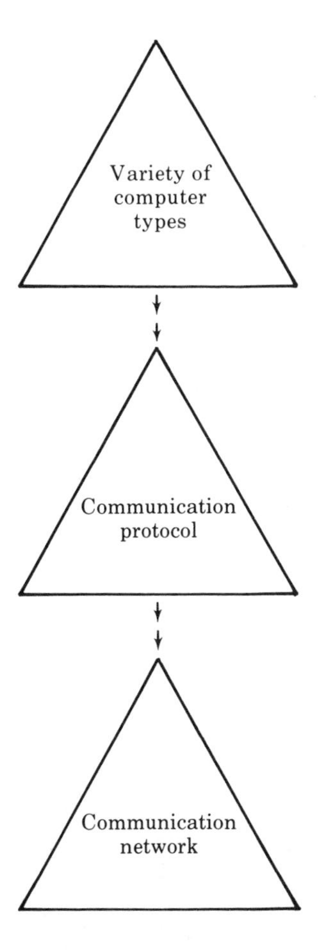

FIGURE 10-7 Costs of computer integration.

controller (PC), microcomputers, numerical control (NC), computer numerical control (CNC) and minicomputers.

The major expense for computer integration within an FMS is due to the number and variety of computers which need to be integrated (Fig. 10-7). Each of the computer controls has its own communication protocol based upon the amount of data that needed to control the component. Thus, the task of computer integration is to establish interfaces and information flow between a wide range of computer types and models.

Before the development of standard communication interfaces, the computer integration required development of software programs

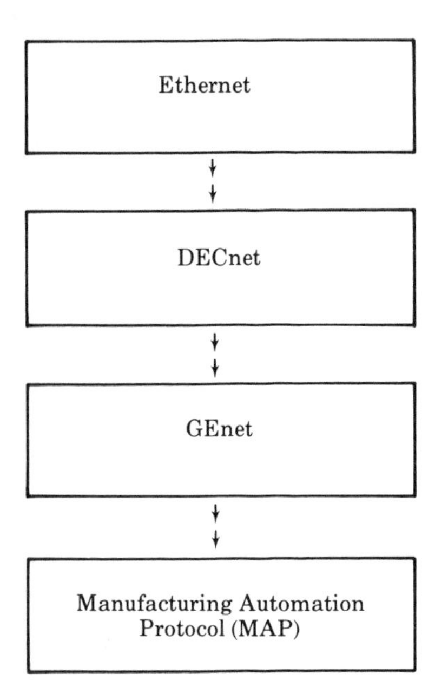

FIGURE 10-8 Standardized communications.

which would handle the information passing between unlike computers. Each new combination required a new development. In these types of installations, the computer integration was hierarchical. This allowed each component to be integrated only with a host computer which would then distribute the necessary information.

Now, with the development of standard communication interfaces (Fig. 10-8), computer control can be integrated directly with unlike controllers. This is arranged through networks. These networks consist of a standard hardware interface such as Ethernet, DECnet (licensed trademark of Digital Equipment Co.), GEnet (licensed trademark of General Electric) or a number of others. Once the hardware can be linked together, the next step toward standard software communication is to permit each controller to send and receive messages over the network. One solution for this problem is the Manufacturing Automation Protocol (MAP). This solution provides standard methods for a PC, CNC, microcomputer or minicomputer to send and receive messages over the network. The development of standard hardware interfaces and MAP has had a significant impact in reducing the cost of computer integration within the FMS. Previously, most of the cost was in time spent getting components to recognize and send correctly formatted data. With MAP and known network hardware, information can be transmitted between computers. Now the cost of computer integration moves to development of the software to utilize the incoming data.

10.3.2 Cost of Software

The computer software provides the ability to transmit timely and accurate status information and to utilize information which has been communicated from other computers. The cost of this software is in development and debugging time (Fig. 10-9). The development and

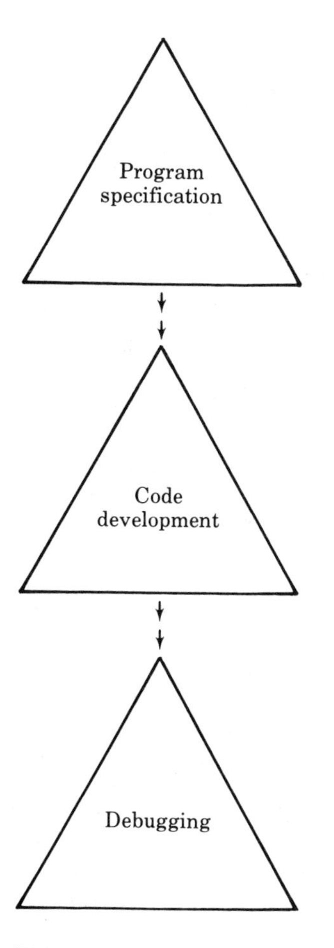

FIGURE 10-9 Cost of software.

cost can be estimated quite well, as it is based upon a specification and the level of programming is fairly simple. It is not nearly as complicated as software development, where a large central database is accessed for all information.

The other cost component to the software is incurred during the debugging process. In many cases, this cost is significant because of the likelihood that it will not be estimated properly. Improper estimates may be due to an inaccurate specification, the trial and error method for debugging an FMS and the fact that testing cannot be done until installation. Chapter 8 describes a procedure by which some of this risk could be reduced with debugging software through simulation. However, the risk of flexible manufacturing is still increased due to the debugging of the integration software. Other aspects of debugging increase the risk of flexible manufacturing as well.

10.3.3 Cost of Operation

One of the fallacies of automation regards the skill level required of the people who must interact with it. One common fear of social scientists in the 1950s was that workers required no skill or education to operate automated machinery. Although this was never proven to be the case, this fear still has a strong relation to current automation projects.

The cost of operating automated machinery might be lower when compared to conventional machinery, but this is not due to the skill level (Fig. 10-10). In fact, the skill level for most automation actually increases, making for savings as the number of necessary workers is reduced. However, savings become questionable when indirect labor costs are considered as well.

Another important aspect of operation costs of automation has to do with equipment failure. When the equipment is highly integrated, the interruptions of one component causes carry-over or integration effects to other components. Therefore, the recovery from such an interruption takes longer when compared to isolated components. The interruption itself might be due to some other integration effect, and a long time might elapse before the actual cause of the problem is found. For example, in a conveyor system which uses pallets to track the parts, a pallet "freezes" in the conveyor. The solution is to simply override an interlock and force the pallet to move through push buttons. However, the initial problem is thought to be a malfunctioning ID switch, and other switches are tested for accuracy. After many tests, it is finally determined that a bolt has retracted out of the pallet which is used in only one instance to signal an end of travel. Because the end of travel is not being signaled, the computer control cannot proceed with the pallet motion.

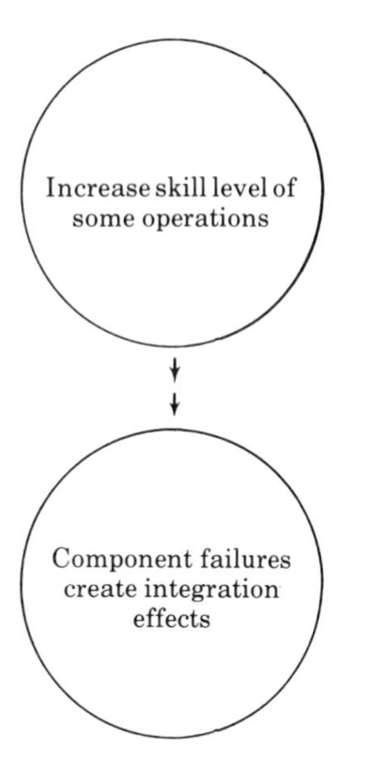

FIGURE 10-10 Cost of operation.

The resolution of this problem requires a period of four days with sporadic interruptions by maintenance to test another possible cause of the problem. But because a method is found to recover from the fault, the entire system is not stopped for repairs.

From these types of experiences, the cost of operating automation is an important component to the overall cost of automation. The other costs which pertain to automation include computer integration, installation, software development and software debugging. These certainly are not the only costs associated with automation, but they have been identified because of their absence in conventional machinery and their common occurrence in flexible manufacturing. But these costs are due entirely to automation, not to the costs associated with flexible manufacturing.

10.4 COST OF FLEXIBLE MANUFACTURING

The costs of flexibility include expenses which are attributed directly to flexibility and those which are due to implementation of the method. Implementation of flexibility can be carried out through various degrees of automation. This section deals only with those costs which pertain specifically to flexibility. Five categories of

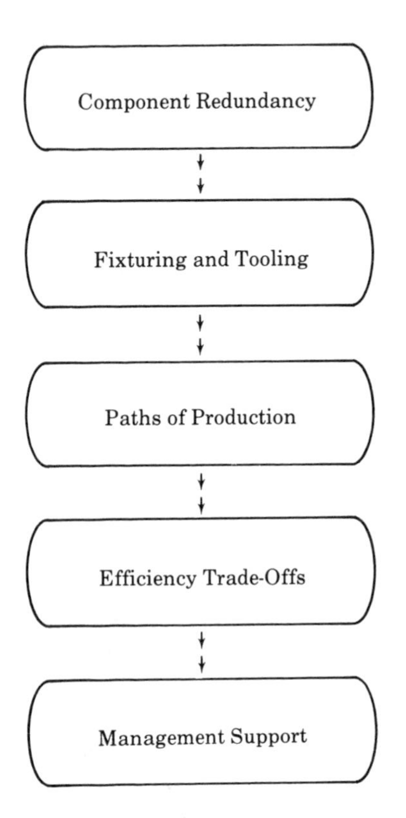

FIGURE 10-11 Cost of flexible manufacturing.

costs have been identified. These include component redundancy, fixturing, paths, efficiency trade-offs and management support (Fig. 10-11). Each is described in detail below.

10.4.1 Component Redundancy

Flexibility is the opportunity for a choice, which exists when there are at least two available options. Continuing this simple logic, flexible manufacturing contains functionally equivalent machinery. This redundancy gives flexible manufacturing systems its name. Functionally redundant components contribute significantly to the cost of flexible manufacturing. For example, it is less expensive to buy one milling machine and one drilling machine, but the more expensive machine which can do either function is more suitable to flexible manufacturing.

For the same production capacity, the components within an FMS will be more expensive for single function type machinery. Another cost of flexibility is when several similar machines are needed but one common machine is selected. This can occur in the number of axes which are present at a machine tool. For example, a given production requirement indicates that one 5-axes machine, two 4-axes machines and two 3-axes machine are needed. To maintain flexibility via component redundancy, five 5-axes machines are required, each one capable of any number of axis operations. This solution increases flexibility, but not without adding some cost through the need for component redundancy.

Once this redundancy is present in flexible manufacturing, the next step is to provide mechanical and control capabilities to take advantage of it. This is accomplished with the use of fixtures and tooling.

10.4.2 Fixturing and Tooling Costs in Flexible Manufacturing

When a single machine to perform some function was acquired, only one function was needed. Now, with the availability of redundant machines, more fixtures are needed, but the number needed cannot be established with a one-for-one calculation.

The primary purpose of the fixturing within an FMS is to provide a uniform orientation of the part so that setups at specific stations do not interrupt the operation of the station. Please note that fixtures do not eliminate the need for setups, but put them up front as part of the installation instead of part of the operation. Therefore, rather than interrupt the operation of a station, flexibility can be taken advantage of at any time without incurring a setup. But this cost has been displaced, not eliminated, through flexibility.

Another characteristic of the cost of flexibility due to fixturing has to do with the number of fixtures. Chapter 2 describes that one way of reducing integration effects is to increase the inventory or work in-process level. This has been applied to the FMS through use of a storage and retrieval system where fixtured parts can be detained until they are needed. In this application, the number of fixtures might be as high as five fixtures per station. This will be a considerable cost of flexibility, yet might eliminate integration effects. A third characteristic of fixturing which affects the cost of flexibility occurs during batch scheduling. Given a production requirement, a single fixture might meet this need on a continuous use basis for the entire year. However, the manufacturing process is batch scheduled and parts will arrive to the FMS in batches, rather than uniformly to their annual production need.

When this situation occurs, the number of necessary fixtures is proportional to the portion of the system's capacity which must be consumed by the batch. For example, if a batch arrives and it is the only part type which is available, sufficient fixturing is needed to consume 100% of the flexible manufacturing system's capacity. If several batches are run simultaneously, the number of fixtures needed will be related to the relative percentage of the system capacity which it must consume.

The number of fixtures contributes to the cost of flexibility just as the redundancy of components does. All of this redundancy increases the price, and also impacts the cost of flexibility in less obvious ways. These impacts relate to installation and quality assurance.

10.4.3 Cost of Alternative Paths in Flexible Manufacturing

A path in flexible manufacturing represents a part sequence of components via a fixture necessary to complete all of its required operations. In a conventional machine environment, only one path exists for a part because a single fixture remains at a single machine. This is not the case within flexible manufacturing systems, where there is more than one path.

To calculate the number of paths within an FMS, the following general formula is presented (Fig. 10-12):

Suppose a part has usable fixtures and that each part requires two operations. There are three alternative stations for the first

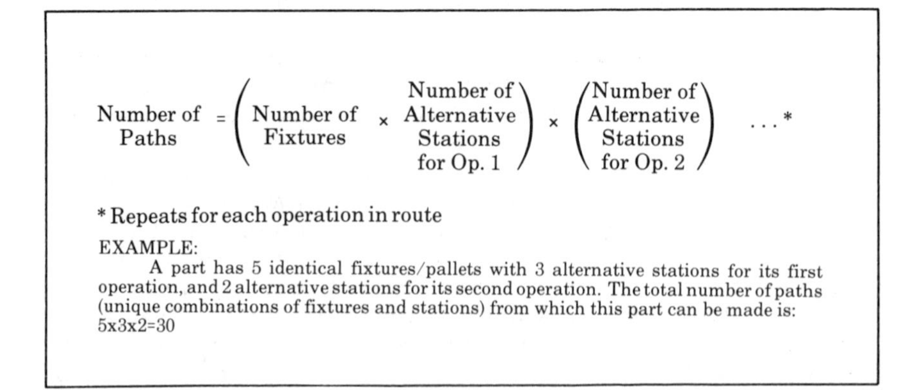

FIGURE 10-12 Paths of production.

operation and two alternative stations for the second operation. In this example, this part would have 5 × 3 × 2 = 30 paths that it could take through the FMS. Thus, instead of having a single path, as might be the case in conventional machinery, the FMS can provide 30 different paths. Of course, each one must be set up during installation to produce the identical part.

The cost for the number of paths is realized in extended installation time and in maintaining the quality of parts for these alternative paths. The number of paths increases the installation time of an FMS. One way of thinking about this is that an FMS moves setup time which would occur during the operation of a conventional machine to the installation. The problem is even greater because the number of setups multiplies with the number of alternatives. Many FMS designs show as much redundancy in components as in fixtures, which results in a large number of paths. These systems might appear to be simple to install but the simplicity of this task will be directly related to the number of paths. Furthermore, after these paths have been set up, they still must be maintained so that they produce identical parts.

This path maintenance is the primary objective of quality assurance within an FMS. The tasks can become extremely large even in the simplest of flexible manufacturing systems. One characteristic of these paths is that random inspection is not suitable, because each path requires a unique setup and if any one of the paths is out of tolerance, poor quality parts will be produced by the FMS, even when a statistical process control system is in place. The quality review must be based upon a round robin testing of each path. This will require a more elaborate management system and will increase the likelihood that poor quality will be observed in many other parts which have potential for the same error.

The number of paths which are present within flexible manufacturing is a measure of the degree of flexibility. But each one adds to the cost as well as through extended installation time and quality review. A means to reduce the cost is to reduce the number of paths by dedicating fewer part varieties to fixtures, or by reducing the number of alternative stations. From the earlier example, it is unlikely that 30 paths are needed for the production of the part. One way to reduce the number of paths is to dedicate three fixtures to a part and to prevent them from sharing generic fixtures. Another way is to eliminate the third station for the first operation. After this change, the number of paths has been reduced from 30 to 12. This is a significant reduction which probably will not affect the production capacity of the FMS, but will significantly reduce its installation and quality review costs.

The dedication of some fixtures or stations to certain parts reduces the number of paths. These act as constraints upon the system which could impact its production capacity through increased integration effects. The relationship between flexibility and operation effectiveness is the fourth area in the cost of flexible manufacturing.

10.4.4 Cost of Operation Effectiveness in Flexible Manufacturing

The operation effectiveness of flexible manufacturing is the matching of appropriate machinery in order to perform a specific operation. The most effective solution is to have dedicated machinery for each operation. This type of equipment will usually have the shortest operation time, but will restrict the number of paths in the FMS. Also, when a dedicated machine has its service interrupted, it will stop the entire production of any part which requires an operation performed at it.

For this reason, more general-purpose equipment is installed in flexible manufacturing. This equipment contains the capability to perform basic functions, such as milling, drilling and boring, with the use of a single tool at a time. The operation times are longer, thus requiring more machine capacity. To obtain the same production capacity in an FMS requires a higher equipment cost than that of conventional machinery. This loss in operation effectiveness is one of the costs incurred for flexible manufacturing.

10.4.5 Cost of Management Support in Flexible Manufacturing

Flexibility adds complexity to the manufacturing process. Some of this complexity is due to the integration effects which occur during its operation, some is due to the number of paths and its related quality assurance task and some is due to the elimination of setup within the process itself. All of these combined present a difficult management problem, which requires a great deal of knowledge and information for effective decision making.

One way to perceive this need for management is through its corresponding role in conventional machinery. With conventional machinery, the strategic decision made by management is to purchase the machine. Operational decisions are then made as part of the shop operation. But in flexible manufacturing systems, the strategic decisions do not end with the installation of the machinery. Flexibility offers the capacity to adapt to a changing environment. These decisions of change, however, require strategic information and are best made by people experienced in strategic decision making. Flexible manufacturing, therefore, requires greater efforts of management support.

Another reason for increased need of management involves the value of each decision's investment and cost. Flexible manufacturing requires a large capital outlay which is paid back over several years. This amount alone warrants the investment of higher levels of management in the operation.

Flexible manufacturing requires indirect management from a variety of areas. One of these is engineering support. Manufacturing engineering is often involved on a day-to-day basis to obtain more efficiency from tooling, or to provide redundancy of operations to bypass stations which have major interruptions in their services. It is common to see some manufacturing engineers working with operators on a regular basis.

Maintenance of flexible manufacturing systems is another area where the need for indirect support is increased. The FMS is comprised of many different components requiring special maintenance supplies and skills. Often, preventative maintenance techniques are implemented to reduce the chance of a major failure which requires more organization and management.

10.5 CONCLUSION

The costs of flexible manufacturing are comprised of the costs of automation, the costs of flexibility and the risk or likelihood that the actual productivity is lower than the planned productivity. All of these costs decrease the return on investment and make flexible manufacturing an expensive solution to a manufacturing need.

There are techniques that will lower the risks or direct costs of flexible manufacturing. As seen in Fig. 10-13, these include limiting the number of paths within the flexible manufacturing system to a realistic level, including some dedicated equipment for

COSTS:	Direct capital investment Risk of FMS Cost of automation Cost of flexible manufacturing
BENEFITS:	Trade-off for inventory Trade-off for balanced loads Adaptable environment

FIGURE 10-13 Cost vs. benefits of FMS.

effective operations, automating where it is appropriate while leaving other tasks to the skills of operators, and computerizing those decisions which can be based upon a limited set of information.

The success of flexible manufacturing depends upon the ability to reduce its risks and costs, but too much avoidance also jeopardizes the project. Flexible manufacturing systems require a large initial capitalization and if this is not accepted, the FMS project will be undercapitalized. The obvious outcome of this is pressure placed upon the installation. This pressure increases the risk as well.

Another way to reduce the cost of flexible manufacturing is to pass the installation risks onto the vendor. The vendor must take on some risks to ensure that an efficient installation is obtained, but the purchaser of a flexible manufacturing system cannot expect the vendor to accept all of the risks. If the risks become too high, the vendor can always pull back his resources. But if all interested parties are not willing to share the risks, it is the purchaser who will be left without a fully operational system and will be unable to manufacture needed parts. Needless to say, it is to his advantage to recognize this fact and enter into a project with a shared risk approach.

11

Acquisition of a Flexible Manufacturing System

Chapters 1–10 have dealt exclusively with the definition of flexible manufacturing. This chapter deals with where and how a flexible manufacturing system can be acquired. The acquisition of conventional manufacturing equipment is quite simple. It starts with a cost-benefit analysis which establishes an ROI (Return on Investment). When the project is funded, vendors are added to prepare a quotation for supplying suitable hardware. One vendor is selected, a contract is signed which provides partial payments for construction, run off before shipment, delivery and installation, and final acceptance run in customers' facility

This method of doing business is well established and effective for both the vendor and customer. The vendor receives payments before actual acceptance, which lowers his risk through less capital investment. The customer controls these payments at strategic points in the business which can be leveraged as needed. Both parties share in the risk of the equipment acquisition, which enhances the chances for a successful project. But these well established, effective procedures for the acquisition of conventional machinery provide only a resemblance to the procedures used in acquiring an FMS. Flexible manufacturing systems change these rules for many reasons. This chapter provides a description of the acquisition of an FMS.

11.1 DEFINITION OF ACQUISITION

Acquisition is the procedure for coming into possession of an FMS. This procedure is described as containing three general milestones

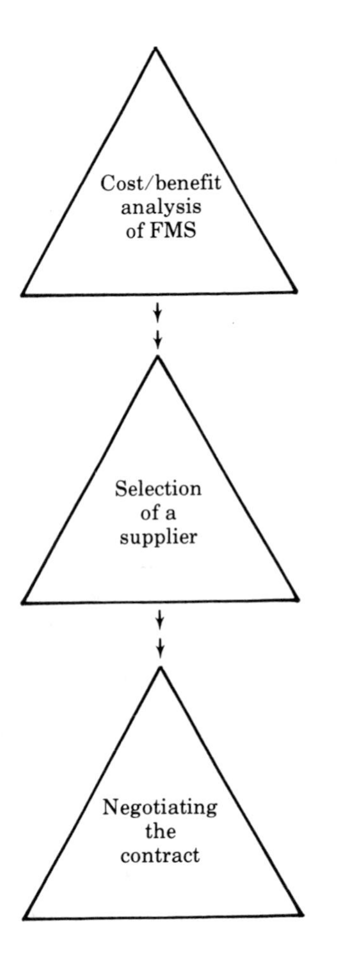

FIGURE 11-1 Milestones of FMS acquisition.

(Fig. 11-1). The first is the decision of what to buy. This decision deals with the alternatives to flexible manufacturing and cost-benefit analyses of manufacturing alternatives. The second milestone in FMS acquisition is the selection of a supplier. This decision involves detailed design, communication and technical know-how. The third milestone is negotiating the contract. It is in this third decision where flexible manufacturing has changed the traditional business methods in equipment acquisition.

The remainder of this chapter addresses each of the three milestones in FMS acquisition.

11.2 FIRST MILESTONE: DECIDING IF AN FMS IS RIGHT

The problem starts with a need to produce a product. This production need is expressed in units for some given period of time which, most likely, is based upon forecasts and judgments about the future business opportunity. But despite this error, the fact remains that there is a production need which must be met.

Meeting the production can be accomplished through many alternatives. These range from outside contracts to completely automated production systems. Somewhere in this diverse range lies flexible manufacturing systems, and the task is to decide if FMS is a viable solution.

This task of FMS suitability is described in four points: the method for economic justification, part selection, process technology and productivity, as illustrated in Fig. 11-2.

11.2.1 Economic Justification Criteria

The decision to purchase an FMS is usually made by top management because of the high capital investment which commonly accompanies

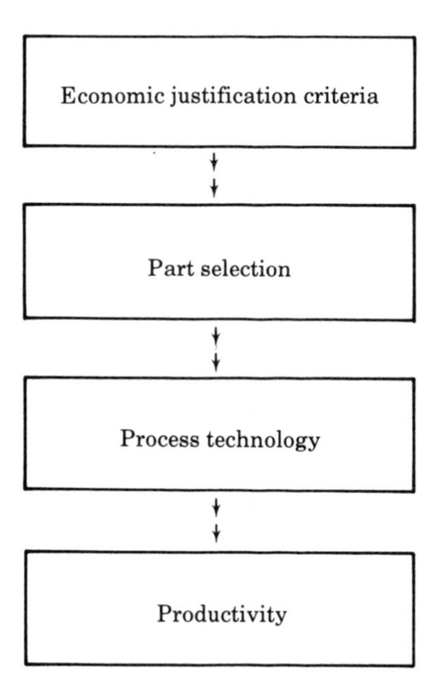

FIGURE 11-2 Cost/benefit analysis of FMS.

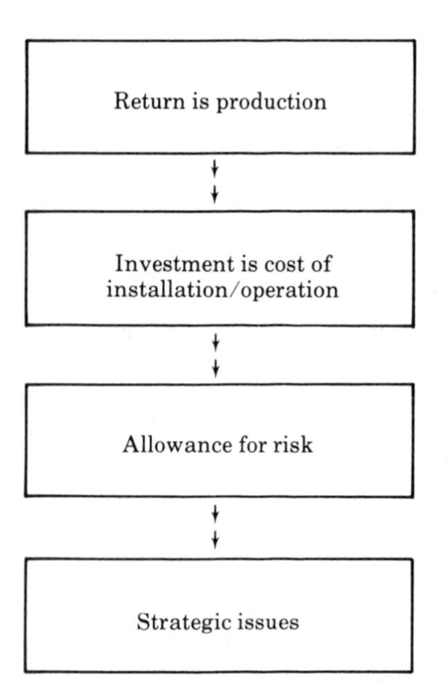

FIGURE 11-3 Return on investment calculation.

the purchase. For this reason, the economic justification is usually based upon more strategic information than cost accounting estimates of ROI. Even in those cases where manufacturing engineers sell the concept to top management by use of ROI calculations, the final decision is more heavily based upon strategic issues rather than on return from an investment.

The reason for this is that flexible manufacturing contains a high degree of risk. This risk can easily shift any ROI, rendering the decision to acquire an FMS based entirely on ROI a gambler's decision (Fig. 11-3). However, none of the gamblers I know run manufacturing companies!

The strategic issues used in deciding whether an FMS is right can be found in the benefits of flexible manufacturing, seen in Fig. 11-4. Chapter 9 contains a description of flexible manufacturing benefits. From this list, the underlying theme of benefits is in the potential production environment which flexible manufacturing can create. This environment permits the traditional long-term decisions to be dealt with in a much shorter time, in some instances in day-to-day operations. For example, the decision to phase out one product line and implement another would traditionally be made

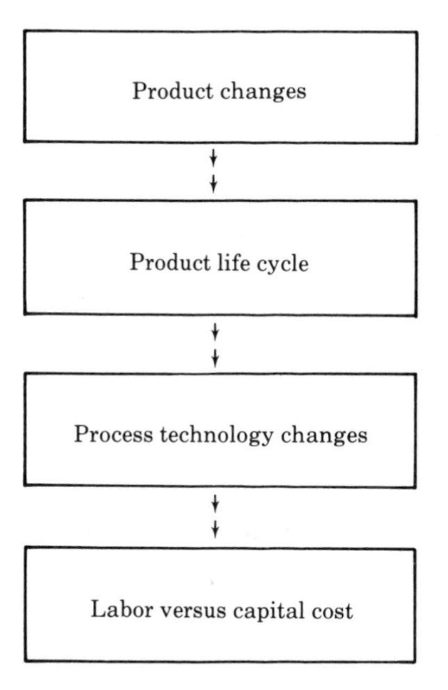

FIGURE 11-4 Strategic issues of FMS.

and carried out over a long period of time. Flexible manufacturing can turn this decision into purely an operational decision where, in a matter of days, the decision can be made and implemented.

The flexible manufacturing production environment affects strategic issues in changing the ratio of cost components in a product. Part costs which are comprised primarily of labor costs can be changed so that the major cost is capital investment. Management might feel variable costs are too unpredictable and might wish to change the costs of components of their product. Flexible manufacturing systems provide a means to do this.

Flexible manufacturing can affect strategic issues relating to implementation of manufacturing technology. The school for learning whether a new technology is applicable is right in the production facility. Management might decide that the ability to understand and successfully implement new technologies into their own production facility is important. Flexible manufacturing systems are one of these technologies; therefore, the decision to install an FMS requires testing the organization's ability to grasp a new idea.

No matter how strategic the acquisition of flexible manufacturing might be, the final decision still comes down to the application and its investment. Strategy will only influence the return in that the potential manufacturing environment will enhance it but the precise amount as in the management instinct. Two items affect the application of a flexible manufacturing system. These are the part and its process.

11.2.2 Part Selection Criteria

Deciding whether or not a flexible manufacturing system is appropriate starts with a production need. All of the strategic advantages will be useless unless the FMS can be installed and operated. It need not reach its expected level of productivity because management has already factored in the return of the potential environment which was created, but it does have to work! One means to increase the likelihood that the FMS will work is to identify suitable parts (Fig. 11-5).

Not all parts can be made in an FMS. The traditional concept of part families suitable to FMS is that they must be of mid-volume production and mid-variety. But this concept provides very little meaningful information in determining whether or not an FMS is suitable for a part family. What really determines the suitability of an FMS is the manufacturing process used in the making of a part.

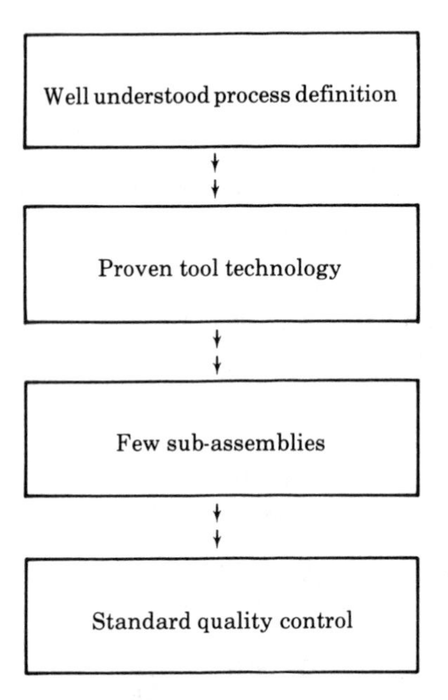

FIGURE 11-5 Characteristics of FMS parts.

11.2.3 Manufacturing Process Criteria

The process, or method, used in the manufacture of a part provides the foundation upon which the FMS is built. Specific requirements of this process exist which indicate whether an FMS will be suitable. These are how well the process is established, its suitability to automation, its suitability to flexibility and the technology of the process itself (Fig. 11-6).

First, the process for the manufacturing of a part must be thoroughly established, tested and implemented before it is suitable for a flexible manufacturing system. Installation of an FMS has sufficient risk already and if justifying a process is added as part of this installation, the risk increases drastically. Too many decisions of an FMS design and its operation are based upon the nature of the processes. When these turn out to be based upon assumptions instead of facts, the flexible manufacturing system's installation is on rocky ground and its risk will increase.

The only process which should be implemented in an FMS is one which is thoroughly understood, tested and proved. The FMS is not the place for exotic tooling and fixturing because its overhead

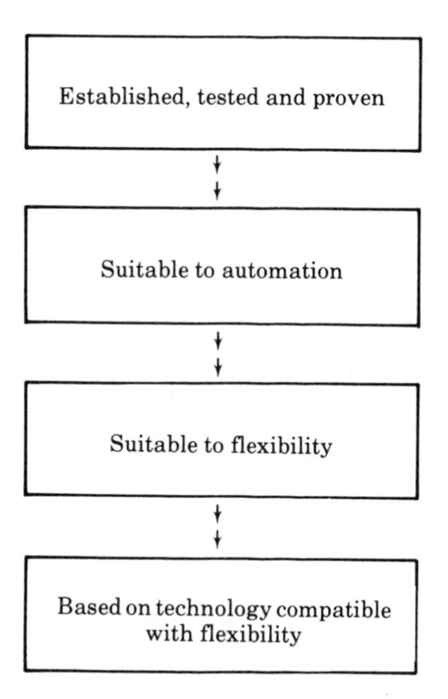

FIGURE 11-6 Process characteristics of FMS parts.

in debugging is too high. Although flexible manufacturing has been used as a means to sell a process, when the process fails, the blame is usually placed on the technology used to implement it rather than on the process itself.

Second, the process must be suitable to automation. This characteristic is needed because most flexible manufacturing systems use a high degree of automation in order to achieve flexibility. Therefore, the process must be able to be implemented via computer controlled equipment and without the need for constant operator attention. If an operation is performed on a part, but the outcome of this operation cannot be identified, it will be difficult to automate. For example, suppose a part requires a surface finish operation but this operation might need to be repeated variously from twice for some parts to five times for others. The number of repeating operations depends upon the operation environment and composition of the parts' raw materials. Establishing an elaborate inspection operation in order to automate this process will result in an increased risk of the FMS and, more likely, its unsuccessful implementation.

For a process to be suitable for automation, it must be currently implemented to the extent where all operation decisions are identified with well-defined sets of solutions. Using John Diebold's concepts, this means that the feedback loops are comprehensively identified. When this is the case, the operation decisions can be programmed into the control and automation will be achieved. But to automate a process where solutions to operation decisions are not identified will cause the application to fail.

Third, the process must be suitable to flexibility. More simply stated, the process must be able to provide more than one path through a production facility. Some redundancy or backup of the operation is essential for the process to be suitable in a flexible manufacturing system. If a process requires one of a kind equipment for every operation, the FMS will not provide any benefits for a conventional solution. Even if one of a kind equipment can be shared by several part families, it is still questionable whether this process is suitable for a flexible manufacturing system. In this case, the flexible manufacturing system can provide flexibility in scheduling, but the benefits from reduced integration effects will not be realized.

The process which is suitable for flexibility is one where operations are defined so that a variety of equipment can be used to perform it. In FMS implementation through use of labor, this means that work stations are designed so that tasks can easily be assigned to one or many operators. In the more mechanical implementation, this means that operations are constructed from a set of individual tasks which can be decoupled and reallocated as needed.

Fourth, the process must contain technologies which are suitable for flexible manufacturing. The technologies used in the process must be suitable to automation and flexibility. This requirement is an indication of why the most common applications of FMS have been in metal removal technologies. The processes of turning, boring, reaming, milling and drilling are fundamentally the same, with each requiring a work table and spindle. The difference between them is primarily accomplished with the tooling. This allows several work stations to have similar capabilities. On the other hand, the process of metal forming is one where several operations are performed at a single station. In this technology, it is more efficient to bend, shape and stamp all operations at a single station where the part is completed. For this technology to be compatible with flexible manufacturing, the individual operation must be decoupled and be able to be performed at different work stations. Some flexibility has been introduced into metal forming technology with the introduction of automated die changing. This provides a potential where operations can be performed for several work stations. But because of the technology of metal forming, only a few flexible manufacturing systems have been applied to these types of operations.

The manufacturing process provides the most information in determining whether flexible manufacturing is suitable. The process must contain operations which can be decoupled into discrete elements, each of which can be shared by the work station. The process must also come from technologies which are fundamentally compatible with automation and flexibility. When a process which does meet these requirements is implemented in flexible manufacturing, it can usually be recognized through the "kill it with hardware" solution. For example, the processing of a part prevented it from being fixtured and held in a fixed orientation for its operations. When this was implemented into an FMS, each part was handled by robots, pick-and-place devices and conveyors over 50 times in its need for 15 operations. There was so much hardware in the material handling system that its collection mean time between failure was five minutes. The failure of the application was not due to flexibility but to the process itself. The successful FMS will be based upon a compatible process.

11.2.4 Productivity Criteria

Determining if flexible manufacturing is suitable relates to the goals of the production environment. These goals might be reduction of work in-process, reduction of direct labor cost, increase in equipment utilization or implementation of current technologies, all of which increase productivity. For flexible manufacturing to be

Goal	FMS
★ Reduce Work In-Process	★ Flexibility is a Substitute for Inventory
★ Reduce Direct Labor Costs	★ Flexibility is Compatible with Automation
★ Increase Equipment Use	★ Flexibility, through Fixturing, Reduces Set-Up Time
★ Use Current Technologies	★ Flexibility Utilizes a Variety of Factory Components

FIGURE 11-7 FMS suitability to goals.

suitable, it must provide benefits to the current goals of the production environment (Fig. 11-7).

If the goal is to reduce work in-process, flexible manufacturing provides the capability to benefit from this goal. Using the Manufacturing Integration Model (MIM) which was presented in Chapter 2, flexibility provides a trade-off for work in-process inventory while controlling integration effects. Therefore, flexible manufacturing provides a means of reducing work in-process.

If the goal is to reduce direct labor costs, automation of the process is required. Flexibility is compatible with automation; in fact, the two are often thought of as one in the same.

If the goal is to increase equipment utilization, one simple solution is to eliminate setups. Flexibility requires that setups be displaced from within the process to parts of the installation. Flexible manufacturing provides the capability for high equipment utilization; in some applications, as high as in transfer lines with perfect balance.

If the goal is to modernize and implement current technologies, flexible manufacturing provides a means to carry these out. A flexible manufacturing system is a technology within itself, but can be made so that pieces of technology can be implemented over a period of time rather than all at once. Chapter 12, which contains a discussion of installation, covers this issue in more detail.

Flexible manufacturing will only be suitable when it meets these four criteria: economic justification, appropriate parts, appropriate processes and compatibility with the goals of the production facility. When all of these criteria are satisfied, flexible manufacturing is right for meeting the production needs. The next step in acquiring an FMS is to select a supplier.

11.3 SELECTING A SUITABLE SUPPLIER

Selecting a suitable supplier is as difficult as determining whether or not flexible manufacturing is suitable in the first place. Three steps are described for deciding upon a supplier. These are preparation of the request for quotation, the proposal/supplier review and the selection criteria.

11.3.1 Preparation for the Request for Quotation

The Request For Quotation (RFQ), summarized in Fig. 11-8, is the document which describes the production need. This is usually expressed in terms of part drawings and forecasted production volumes. In some instances, the process is specifically spelled out in the RFQ as well, but usually this is left out, giving the supplier an opportunity for creativity.

Accompanying this description of the production need are the objectives of the implementation. These reflect the overall goals such as reduction in direct labor or work in-process inventory level. It is important to state these objectives because they will be used in the selection criteria in establishing how well a potential supplier is able to address the specific production need.

The third component of the RFQ is a description of the production environment where flexible manufacturing will be implemented. This includes a description of where, when and how parts reach the point in the process, and where, when and how they are expected to leave the process. This also includes a description of product structures with assemblies, subcomponents and raw materials. Furthermore, it includes a description of the degree of automation and flexibility in other processes which affect the one under consideration.

When any information is omitted regarding the production need, goals or the production equipment, the supplier will make assumptions. These assumptions will have important impact upon the flexible manufacturing system proposal and often will not be explained or listed. In some instances, these assumptions about the production environment remain until errors are discovered during actual installation. By that time, however, substantial costs are usually involved in correcting these omissions. The best means to prevent an FMS design based upon assumption is to provide these three items of information as part of the request for quotation.

11.3.2 Proposal/Supplier Review

As a result of the RFQ, each potential supplier will prepare a proposal. Some proposals might be only conceptual where others might

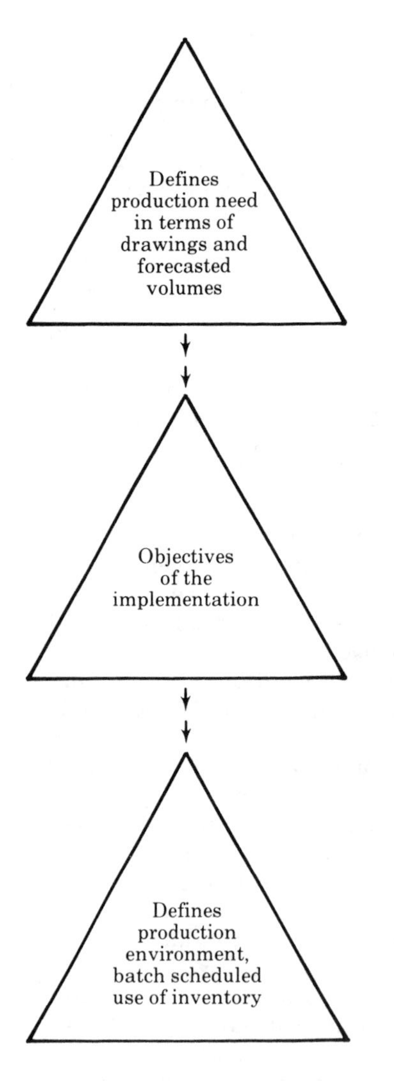

FIGURE 11-8 RFQ contents.

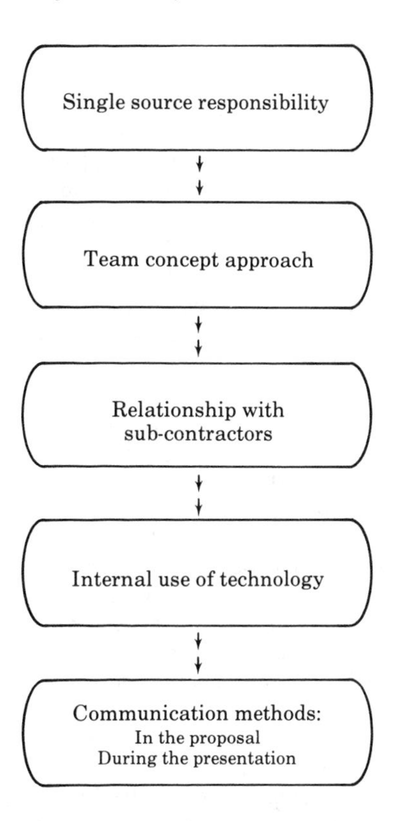

FIGURE 11-9 Characteristics for supplier review.

contain thorough detail. No matter how much detail is included, the proposal will contain errors.

The review of these proposals is not to identify the flaws but to identify a supplier who has the potential to help to pinpoint these errors and is willing and able to correct them. There are several characteristics which identify such a supplier (Fig. 11-9).

The first characteristic is single source responsibility. Due to the diversity of components within a flexible manufacturing system, no supplier will manufacture all of the components. Rather, a collection of components from a variety of vendors will be collected by the supplier. Single source responsibility means that the supplier can get what is needed from individual vendors throughout the FMS project. This does not mean that the supplier must own or have financial interest in each vendor, but means that he can demonstrate their working relationship and mutual interest in the success of the project.

No single supplier will have the latest technology for the flexible manufacturing application. Instead, suppliers must demonstrate that they can find the best technology and bring it to the project. For this reason, a team approach to supplying a flexible manufacturing system has been appropriate in some projects. The team is comprised of a group of companies, many of which are vendors, each with expertise in a specific area.

In the evaluation of a team, the review must focus on team organization, how communication will be handled, how decisions will be reached and how responsibility will be shared or will fall upon the team organizer. Teams can be effective for bringing expertise and technology into an FMS project, but all of this talent must be well organized for efficient operation. The single source supplier has the advantage of organization, but might lack the freedom to use appropriate technology because of competition.

Another revealing characteristic is the potential supplier's own use of current technology. An obvious measure of this can be found in the supplier's presentation of the proposal. A proposal which contains over 400 pages of written material demonstrates a much different awareness than a proposal which contains some computer graphics, a videotape or even a color graphics animation of the flexible manufacturing concept.

The supplier's own business operation also gauges an ability to understand and implement technology. If the supplier operates a business which excludes the application of flexible manufacturing, this indicates a weak attempt to understand it. On the other hand, when the supplier demonstrates an implementation of technology into business, this indicates positive attitude to challenge. A flexible manufacturing system will always be a challenge!

Another identifying characteristic of the supplier is an ability to recognize errors and to work for solutions. This characteristic is most evident during the oral presentation. When the FMS concept is being presented by the supplier, the need for clarification of some points and even mistakes will arise. How the supplier responds to these is an indication of response throughout the FMS project. When a question is raised or an error is identified and the supplier performs some tabletop engineering to find an immediate solution, usually the first idea becomes the solution. Further investigation is always needed in these cases, but when the supplier is more interested in any solution than *the* solution, this indicates a tendency toward finishing the job without careful scrutiny of all possible solutions.

Probably the most important characteristic of a flexible manufacturing supplier is an ability to adapt a concept into a realistic solution. This requires that the supplier have access to expertise as needed, both inside and outside of their own company. It also means that the supplier believes in technology and understands that it requires special input for successful implementation. Adapting a concept requires that a supplier be able to respond to questions by first recognizing that there is a problem and then by taking the time needed to provide an effective solution. A supplier with these characteristics can be considered as a candidate for the flexible manufacturing project.

11.3.3 Selection Criteria

The selected flexible manufacturing supplier will be directly related to the success of the project. A poor choice will result in a poor FMS and a good choice will increase the likelihood that the FMS will work. Five items are listed below and in Fig. 11-10 which help in determining the appropriate supplier.

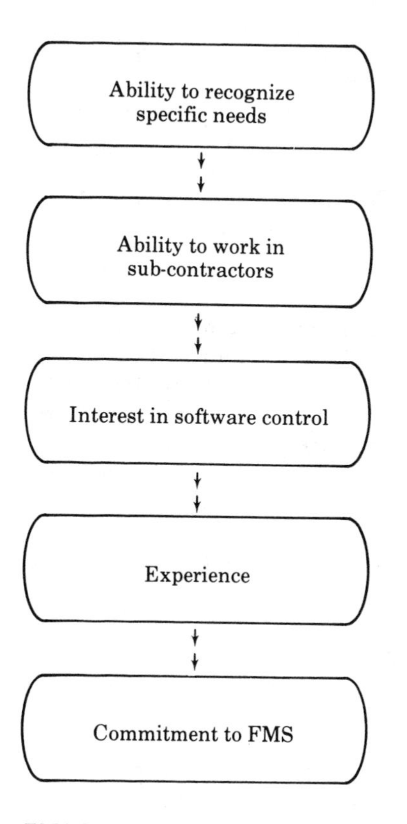

FIGURE 11-10 Selection of an FMS supplier.

Ability to Recognize Specific Needs

The most important criterion in selecting a supplier is an ability to recognize the critical problem. Some suppliers might suggest that the problem is material handling, or even scheduling control. In any case, the solution which is offered by the supplier must be compatible with the customer's perception of the problem. If it is not, the supplier should not be selected; the supplier is pushing a solution rather than taking the time to recognize and define a problem and then construct an effective solution.

Ability to Work with Subcontractors

The supplier's primary responsibility is to recognize the specific needs of the flexible manufacturing project and find effective solutions. This will require expertise from outside sources, so the supplier must be able to get the necessary information and products from vendors. Therefore, the second selection criterion of the flexible manufacturing supplier is an ability to identify vendors, know their capabilities and to establish a good business relationship with them. Some suppliers might indicate that they provide single source responsibility and there is no need for outside vendors. But this is an overstatement of the fact that no single organization can possibly provide expertise in every component, control or other component of a flexible manufacturing system. The flexible manufacturing project will involve a collection of vendors and the supplier must be able to organize and tap the resource as it is needed.

Interest in Software Control

Most flexible manufacturing suppliers have backgrounds as either individual component vendors or transfer line vendors. Both of these involve the vending of hardware solutions to production problems. The FMS offers not only hardware but also a complete control capability. It is important for the supplier to have an interest and understanding of the control side of flexible manufacturing, although previous experience may have only dealt with hardware.

Experience

There are a lot of flexible manufacturing suppliers who have never installed a flexible manufacturing system. And after the first installation, most take a very serious look at whether they wish to continue in this business or not. After the review, some decide the risk is not worth the potential profit and drop out of active pursuit of flexible manufacturing system projects.

No matter how difficult the first installation was, the flexible manufacturing system supplier is more knowledgeable about the hardware and its ability to operate in the computer integrated environment. The supplier's proposals will usually reflect this. Often, the more experienced vendor will have a proposal which appears to be more conservative than a less experienced supplier. All suppliers underestimate the risk associated with the flexible manufacturing system installation and find they can eliminate some of this risk through simpler, less technical design. Do not be misled to think that state of the art technology has its home in flexible manufacturing systems! There is a great deal of experience required to implement a new technology and flexible manufacturing is not usually the most cost-effective place to learn.

Commitment to Flexible Manufacturing Systems

The fifth criterion for selection of a flexible manufacturing supplier is a commitment to flexible manufacturing systems. If the supplier is interested in jumping on the bandwagon and riding the automation as far as possible, when the going gets tough, the vendor will fall back to a safe position. But the supplier who is committed to remaining in the flexible manufacturing business will be more interested in seeing the FMS project become successful. These projects are too expensive, require too much expertise and last too long for any quick profit operation. Any organization which is sincerely interested in flexible manufacturing has to be committed for the long run and be able to accept some of the risk which is associated with an FMS.

The selection of a flexible manufacturing supplier starts at preparation of the RFQ and continues through proposal review and evaluation to ranking each supplier according to some criteria. As this business relation is developed, the people involved hopefully acquire a good feeling about each other's abilities and sincerity in seeing the project through successfully. If any more initial doubt develops, the actual flexible manufacturing system installation will only provide more doubt and, consequently, confidence in abilities will decline. For the FMS project to be successful, this situation should be avoided. Upon selection of the supplier, the first formal step in initiating the business is to negotiate the contract.

11.4 NEGOTIATING THE CONTRACT

There are three components to a flexible manufacturing acquisition contract. These are statement of responsibilities, who accepts the risk and the acceptance of criteria. Each of these is described as follows.

11.4.1 Responsibilities

The statement of responsibilities of the flexible manufacturing supplier, seen in Fig. 11-11, includes proof of design, delivery, installation procedures and start-up of the systems operation. The contract should contain the specific methods by which the supplier will address these issues. Along with these points are the schedule and cost.

These items are common in almost any contract where production machinery is delivered. Because flexible manufacturing systems are a new technology within themselves, the customer has responsibilities as well. The customer cannot expect a supplier to spend nearly two years working with a flexible manufacturing system design, installation and start-up and pass all of this knowledge along in a 30-day runoff period. The customer has the responsibility to learn and understand as much as possible about the FMS being received. This education must take place throughout the installment period because there will not be time for learning when the system begins operation. At that time, only response to situations will be important. And without sufficient knowledge about the flexible manufacturing system, these situation responses make for a very flat learning curve.

One effective tool in learning the FMS prior to its installation is the use of computer simulation. This tool can provide a system level view which is the most effective level for learning. Once the system view is obtained, people can deal with a specific situation and be able to place it in the system.

Responsibilities Of Supplier	Responsibilities of Customer
★ Detailed Design	★ Thoroughly Evaluate Design
★ Proof of Design (Performance and Technology)	★ Accept Proof of Design
★ Delivery and Component Acceptance Tests	★ Learn from Supplier
★ Installation Procedures	★ Understand the Integration Effects Inherent in System
★ Start-Up of System Operation	★ Share Risk of the Project

FIGURE 11-11 FMS contract responsibilities.

11.4.2 Sharing Risks

One of the important costs of flexible manufacturing systems is the risk associated with its installation. There is a high likelihood that the installation will take longer and cost more than expected. This will cut into any return on investment calculations and turn economic feasibility into infeasibility.

Many flexible manufacturing system buyers have recognized this risk and have attempted to deal with it in the contract. The way in which they choose to deal with it is to pass on as much risk as possible to the supplier. This is evident by the presence of many contractual clauses (see Fig. 11-12).

One way of passing risk to the supplier is through the payment schedule. The purchaser might propose that no payment be made on any piece of equipment until it has passed a partial acceptance test. For these situations, the flexible manufacturing supplier is financing the building of the equipment.

Another way of transferring risk to the supplier is through delivery date and penalty clauses. These indicate that if the FMS is not operating by a specific date, the supplier will be requested to pay penalties for each day the purchaser is without production capacity.

A third means for transferring risk is to require a fixed cost project. This means that any improvements or knowledge gained during the actual project will only be implemented if it saves or has no effect upon the cost. In these arrangements, good ideas are often discarded because the supplier loses interest in effectiveness and focuses only on the cost of the project.

A fourth method for transferring risk is through the acceptance criterion. This is the subject of the following section.

The impact of pushing more risk onto the supplier will only be negative on the FMS project. No matter how desirable a low risk contract is, the purchaser is the eventual loser if the flexible manufacturing system does not work. Perhaps the purchaser will be able to save some money or cause the bankruptcy of the supplier, but the buyer will still be left without a solution to meeting a production need. This is where the real cost comes in.

The acquirer of flexible manufacturing must have a shared interest in seeing the FMS project succeed. The only way to demonstrate this is to negotiate a contract where the risk is shared. When a new solution is discovered, the cost of it is shared by both the supplier and customer. When a due date passes, revised schedules are agreed upon by both parties. When unexpected situations arise, a compromise is the best method to ensure a successful project. This is accomplished by sharing the risks of the FMS installation.

If either the customer or supplier is able to create an imbalance in the risk, this in itself increases the risk of the flexible manufacturing project. These projects are already more at risk than many companies can accept, but attempts to avoid some risk through the contract only increases the risk of the entire project. The eventual loser when a project fails is the customer, who is left with no means to manufacture a product.

11.4.3 Acceptance Criterion

The occurrence of fixed cost projects, excessive penalty clauses and unreasonable payment plans have found places in flexible manufacturing contracts. All of these are new and different from the contracts for conventional machine acquisition. But the area which has become the most creative in FMS contracts is the acceptance criterion, illustrated in Fig. 11-12.

The acceptance criterion is the measurement which will be made upon completion of the installation. If this measurement meets or exceeds the level defined in the contract, the customer accepts the system and makes the final payment to the supplier. These measurements of acceptance have ranged from parts per hour to system uptime.

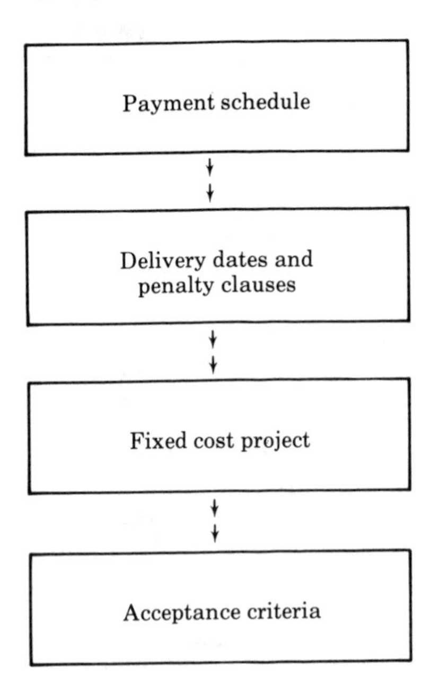

FIGURE 11-12 Shared risks.

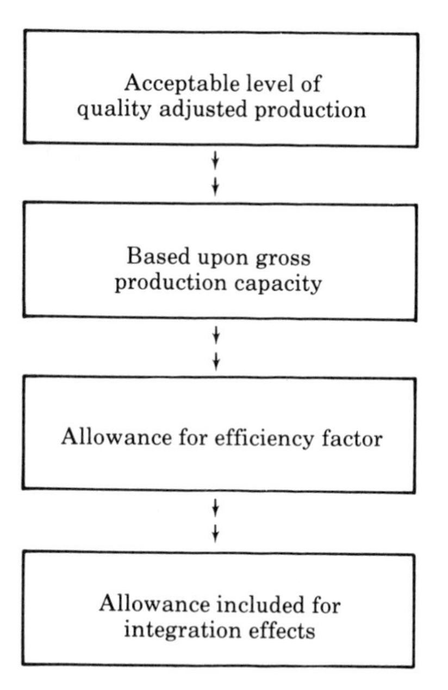

FIGURE 11-13 Acceptance criterion.

The traditional measurement for acceptance criteria is parts per hour. In the case of acquisition of a single work station, this number is quite easy to observe. When the station fails and interrupts, this rate is understood and as long as it can be repaired so as not to fall below its efficiency level, the station will achieve a desired level of production. The component will be accepted and the contract will conclude successfully.

However, this is not so simple in the FMS. The system as a whole can achieve some production rate. But due to its size and complexity, it is impossible to obtain a long enough period of operation to determine its 100% efficiency production level. It is normal for a component to fail and for the FMS to still be able to continue. But integration effects within the system prevent any simple estimate of uptime and production rate. For this reason, some exotic acceptance criteria have been proposed.

An acceptance criterion which will really work must contain some allowances for the integration effects which will occur during the operation of the FMS. It cannot be expected that the FMS can operate for several shifts without a component failure, nor can it be expected that integration effects can be avoided. The decision to acquire flexible manufacturing includes the knowledge that integration effects will be present in the production facility. Therefore, an acceptance criterion must recognize the characteristics of the FMS operation.

One way in which this has been implemented in an acceptance criterion is through the use of computer simulation. In these applications, the FMS is operated for the acceptance time period. Then, a computer simulation is run for the same time period where there are no component failures, a random pattern of failures and an equal pattern of failures. The results of the computer simulations are compared to the actual results to identify the system operation. If the system's actual performance equals or exceeds those of the simulation, the system is accepted.

This type of acceptance criterion still contains some difficult issues, such as which simulation is used, who does the simulation and how accurate the results must be between actual and simulated for acceptance. But despite these shortcomings, this acceptance criterion at least represents a feasible solution to terminate contracts, many of which remain unresolved today.

11.5 CONCLUSION

The acquisition of flexible manufacturing begins with the decision to produce a part and ends with the negotiation of a contract to construct a facility to produce the part. In order for this facility to be an FMS, specific characteristics must be present. First, the part must contain a process for manufacture which is compatible with flexible manufacturing. This means that alternatives must exist for fixturing, work stations, tooling or material handling. If a process requires too much dedication of equipment, only inventory can be used to manage the integration effects.

The application of flexible manufacturing to a dedicated type process will not provide effective management of integration effects and productivity will be lower than that of a conventional manufacturing facility.

Given that the process is suitable for flexible manufacturing, the next step is to find a suitable supplier. This supplier can be comprised of a team of vendors or one company representing several vendors. In either case, the organization and management structure are important in identifying whether the supplier can oversee a complicated and risky project as an FMS.

After the review and selection of a suitable supplier, the next step is to negotiate the contract. This contract defines the responsibilities of both supplier and customer and should not be used as a means to displace risk from the customer to the supplier. If the project is perceived as risky and the customer is not willing to share in this risk, the contract is not the place to shift responsibility to the supplier.

The contract must provide a means where risk of the project is equally shared between customer and supplier. It should address delays and cost overruns, but not through outrageous penalty clauses. It must be recognized that the customer is the eventual loser if a production need is not met, and flexibility offers a great deal of potential, much of which cannot be quantified until those opportunities exist.

The last point in negotiating a contract is the acceptance criterion. Acceptance, based upon some target rate of production, is not realistic for flexible manufacturing. Any rate of production must be qualified with a given level of component availability. Because of this need to qualify the performance, acceptance criterion can be based upon a computer simulation. The computer simulation is run with as equivalent a loss in component availability as in the real system. The actual production can then be compared to the results of the simulation to define whether integration effects are being dealt with properly in the real system.

Finally, the choice of flexible manufacturing offers a new potential for manufacturing. But this opportunity does not come without some risk. The customer must be willing to share in this risk. When a contract is used to avoid this risk, the chances for a successful FMS are significantly reduced.

12

Installation and Operation of Flexible Manufacturing Systems

The decision to purchase an FMS is based upon a design which yields an expected opportunity of production that is higher than its respective risk. The decision to acquire an FMS is usually made from a conceptual design with the detailed design to follow as one of the first steps of the delivery and installation process. This conceptual design has undergone close evaluation, yet several decisions are based upon assumptions and expectations. These must not be forgotten or ignored as the details of the design become known.

12.1 INTRODUCTION

Close monitoring of the detailed design is one means to reduce the risk of the FMS. The transition of assumption to fact requires expertise, most of which must come from the customer. If the vendors were more knowledgeable about the manufacturing of a product than the customer, they would be making the product and not selling technology to produce it!

As details of the design are uncovered, each must be identified as either conforming to or violating any assumption included in the conceptual design. These corrections provide the means to ensure that the FMS project maintains a uniform level of expectation between vendor and customer. When the customer does not follow these corrections to design, his expectations of the FMS will be different than those of the vendor. When this happens, the vendor can proceed with an installation that achieves their perception of the design goals but may not meet those of the customer. At this point, the primary problem is not with the actual FMS but with attempting to compromise between the expectation of the customer and what the vendor actually installed. The energy spent on the resolution of this conflict wastes resources and adds more pressure to an already risky project.

The customer has a primary responsibility during the installation, which is to provide resources necessary to maintain a constant awareness of the detailed design and installation procedure. This involves a constant review of assumption and noting whether the factual design conforms to these assumptions. This monitoring also provides an effective means for transferring knowledge between vendor and customer.

The transfer of knowledge is an important aspect of the installation. But how this transfer takes place depends upon the method used for the FMS installation. These methods are phased in, single source and self contracted, each of which are described in the following section.

The remainder of this chapter contains a description of the procedure for installing an FMS. To illustrate this procedure, a case study of photographs are used. These photographs were taken from several different installations. Special appreciation is given to Gidding & Lewis Corporation who donated the majority of the philosophy for this chapter. This series of pictures provides a chronological installation of the FMS.

The final section describes some general characteristics of the operation of the FMS. One of the most difficult stages within an FMS installation is the transition from a vendor operated system to a customer operated system. This transfer involves selection, training, defined roles and responsibilities. On top of this must be an incentive system from which operators will be motivated to carry out their respective responsibilities.

12.2 METHODS OF INSTALLATION

The methods of installation include phased in over a period of time, single source responsibility and self contractor, seen in Fig. 12-1, where the customers themselves perform contractor roles. In each method, the number of responsibilities remain the same; however, they are allocated between supplier and customer in different manners. This displacement of responsibility is further described for each of the three methods of installation.

12.3 PHASED IN INSTALLATION

Phased in installation of an FMS occurs whenever an FMS is installed with the capacity to meet an interim need with additional capacity

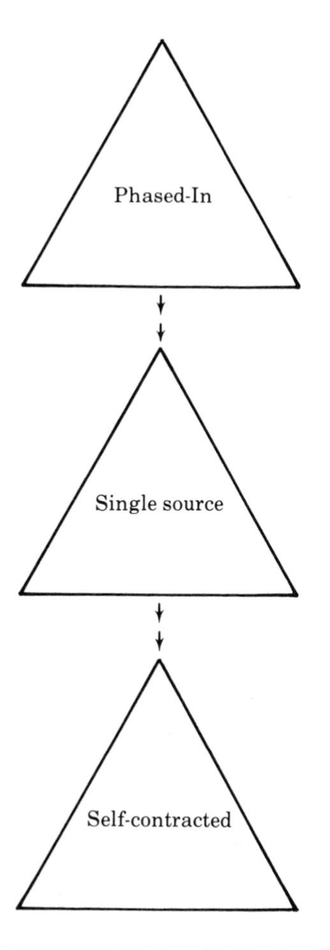

FIGURE 12-1 Methods for FMS installation.

obtained as the forecasted need increases. This acquisition of capacity is not done on a continuous basis but must be obtained in discrete amounts. The frequency and amount of obtained capacity is determined from the forecasted growth in production need and capital budget plans. An illustration of phased in acquisition as it relates to production need is shown in Fig. 12-2.

This strategy of acquiring production capacity at strategic intervals has economic appeal. However, what is fasible and what is practical must be thoroughly understood.

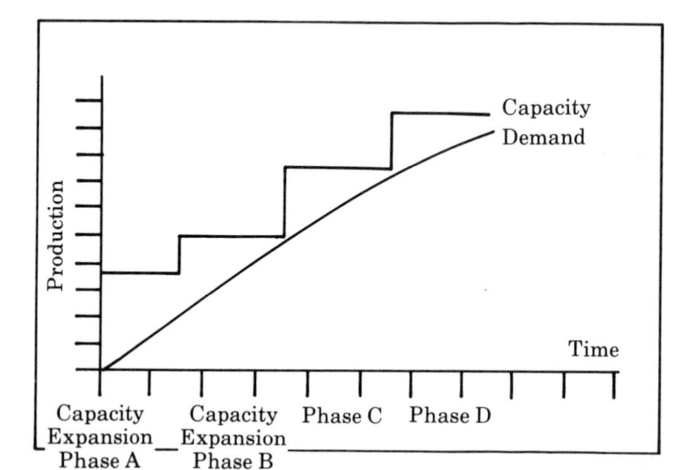

FIGURE 12-2 Phased in capacity expansion.

12.3.1 What is Feasible?

The first point of feasible acquisition of FMS capacity is that the capacity comes in integer amounts. This integer capacity reflects that whole machines, pallets and transporters are added and that installation of additional capacity requires that the FMS be available for integration of the new components. This interruption of FMS production has a discrete impact upon the production of the FMS. The optimal solution is to add production capacity to track with forecasted production. But this type of continuous acquisition is not feasible because of the nature of integrated production systems. Feasible phased in installation of FMS requires that discrete amounts of capacity be added at strategic periods of time. Planning of such additions must be done well in advance of the actual installation.

12.3.2 What is Practical?

What can feasibly be done in phased in installation of an FMS versus what is practical are quite different things! On the practical side, the capital expense savings of phased in installation are not sufficient to offset the cost in terms of lost production resulting from the interruption in the FMS operation.

Phased in installation is practical with production systems which contain a low level of integration. However, as the level of integration increases, it becomes less practical. In the case of FMS, phased in installation has proven to be impractical due to the interruptions in production that it causes. Several flexible manufacturing systems were installed with the intention of increasing capacity in a few years but very few have actually followed through with this plan.

Some plans called for integration debugging during the third shift and on weekends, or during other nonproduction periods of time. The time required for preparing the system for testing and ensuring that it is ready for production leaves only a few hours of a shift for testing. In such a case, the debugging of the integration will require many months. Even if some of it is done during a shutdown period, the debugging time will extend into the production period.

A solution to this problem of system unavailability for debugging can be resolved by stopping production on the system for a period of one month (Fig. 12-3). Even this period of time is too short, but in the case where parts are subcontracted for one month, it is necessary to build one month of inventory for the FMS to be taken out of production for that length of time. Both of these solutions are very expensive, which makes them impractical.

What is practical for phased in installation is the reduction of FMS capacity to a semi-manual level where debugging can be done concurrently with production (Fig. 12-4). In this situation, operators are needed at each station to begin and monitor the operation. They, in turn, signal when a part can be routed and decide to which station it must travel. These instructions are sent to the material handling system, which accepts mutual requests. In this environment, specific pallets can be removed from production for testing at

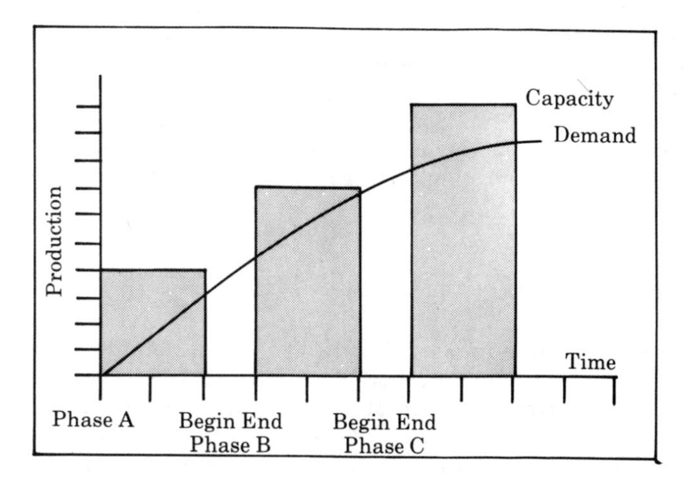

FIGURE 12-3 Phased in installation with system shut down.

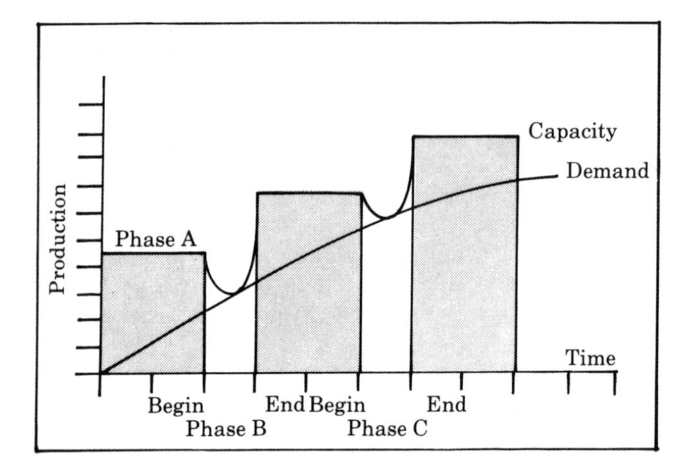

FIGURE 12-4 Phased in installation with semimanual operation.

the new stations. At specific times, an original station can be taken out of production for the testing of new pallets. This iterative process of matching new pallets to original stations and original pallets to new stations can proceed while the system continues to operate in a semiautomatic mode.

The operator in semiautomatic mode will reduce the production capacity; therefore, the addition of capacity to an FMS must be done during a period when either an inventory has been built or when the reduced level of capacity is sufficient to meet production needs. The occurrence of either of these situations of inventory or excess capacity is quite unlikely and, as a result, the cost due to interruptions in production makes FMS expansion impractical.

The phased in installation has intuitive appeal, but when the integration is considered from a realistic point of view, it is not very practical. For this reason, the single source type of installation has been the most common.

12.4 SINGLE SOURCE INSTALLATION

Single source can take on two forms (Fig. 12-5). First, single source can mean that all components, both hardware and software, are constructed by one company. The other type of single source is where one company is responsible for all components. In this second form, the vendor can choose suitable subcontractors as they are needed for specific projects. The advantage of the single vendor

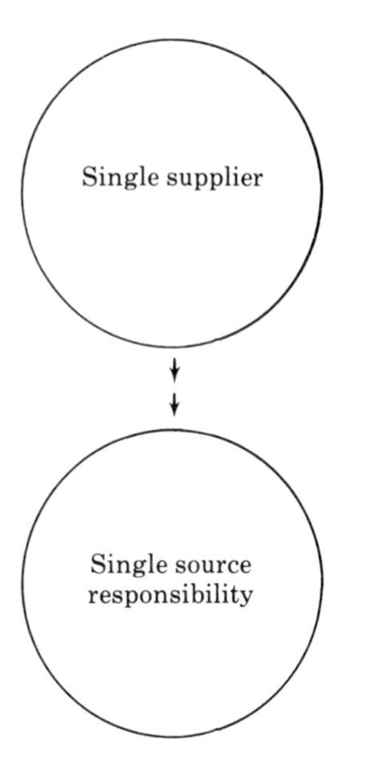

FIGURE 12-5 Single source installation.

type of single source is that resources can be allocated but there is no choice from various alternative sources. The single source responsibility form has the advantage of selecting vendors as they are needed, but coordination of their activities requires an organized management system.

Both forms of single source are used, each having its strengths and weaknesses. One method cannot be recommended over the other. Suffice it to say that, regardless of which single source method is chosen, it will result in the completion of the FMS installation.

12.4.1 Turn Key Approach

A popular phrase that accompanies single source installation is *turn key*, meaning that the vendor is entirely responsible for the installation and start-up of the FMS. The customer does not need to provide any resources until the first day of operation.

The turn key approach is not desirable for an FMS and can even increase the risk of installation. Flexible manufacturing systems are production systems which never reach a constant mode of operation. They are dynamic due to the nature of flexibility. Learning and understanding this dynamic nature of FMS is an important step in bringing it to a production environment. The turn key approach prevents the customer from becoming actively involved, having no opportunity to learn of the dynamic characteristics. An important part of any FMS installation is the transfer of knowledge from vendor to customer. A turn key approach does not provide an environment where the transfer of knowledge can occur.

12.4.2 Transfer of Knowledge

One of the most important outcomes of the installation should be the transfer of knowledge from vendor to customer (Fig. 12-6). The vendor knows the strengths and weaknesses of each component and the customer must obtain this knowledge.

Due to the trial and error procedure used in debugging the FMS installation, customers prefer that their operators not be exposed to the system for fear of affecting worker morale. Operators will have high expectations, but when a problem occurs which requires some debugging, their expectations and morale will drop. One way to avoid this drop in operator morale is to keep them out of the trial and error installation period. In these cases, the transfer of knowledge is obtained through classroom training, which usually only addresses the "normal" mode of operation. When the operator joins the FMS, he soon recognizes that the normal mode of operation is seldom achieved and morale drops anyway.

Worker morale is important to consider, but the need to understand the FMS is of much greater importance. The installation provides an opportunity where knowledge can be transferred and operator involvement should be looked at as knowledge gathering. This does not mean that operators need be involved with installation continuously, but regular involvement provides a means for knowledge to transfer from vendor to customer.

12.4.3 Responsibilities of Vendor and Customer

In the single source installation, the vendor has most of the responsibility. These responsibilities, described in Fig. 12-7, include construction, preparation of the site, delivery and installation of components, integration of hardware, integration of computer control, communication interfaces, debugging, testing and runoff of acceptable parts. These tasks will require a variety of skills, each of which must be arranged for and acquired at appropriate times during the installation. The most important responsibility of the vendor is to organize and obtain the appropriate resources as they are needed during the installation.

The customer has fewer responsibilities in a single source installation. These responsibilities include allocation of floor space, provision of support sources such as electricity, heating, offices,

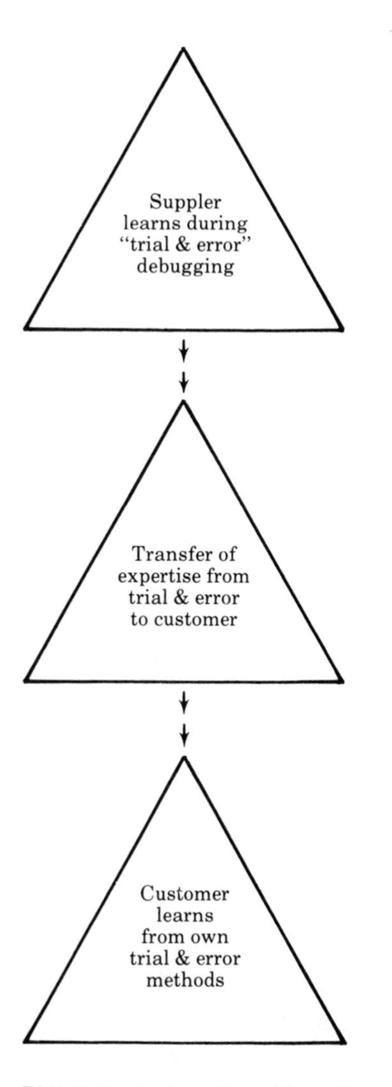

FIGURE 12-6 Transfer of knowledge.

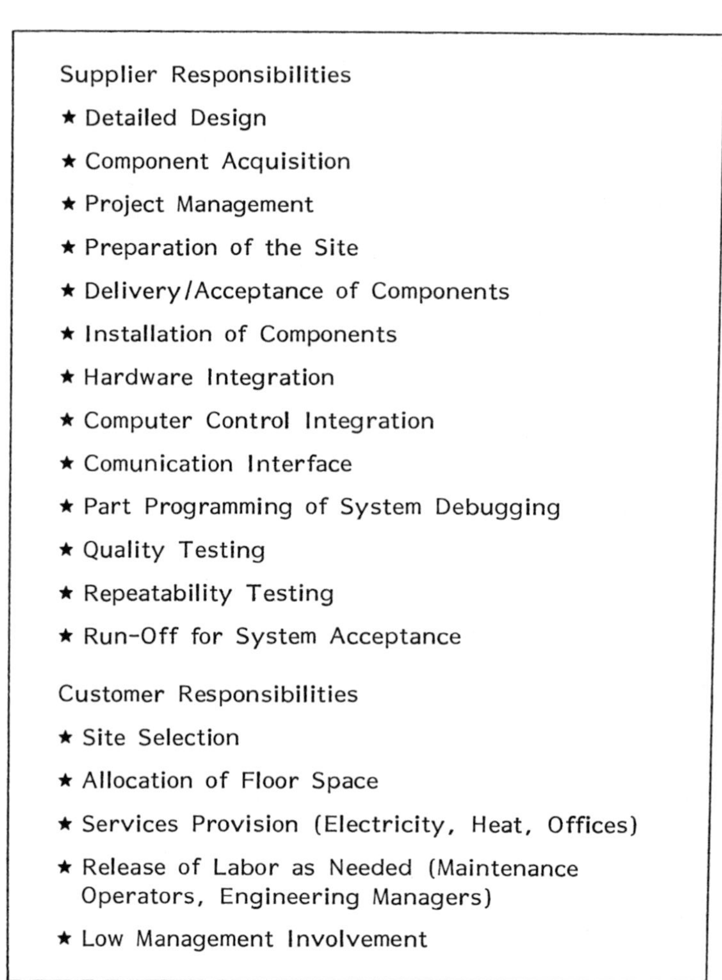

FIGURE 12-7 Single source type installation.

progress payments and release of workers as they might be needed in the installation. Usually, very little management or coordination of these responsibilities is needed.

12.5 SELF-CONTRACTED INSTALLATION

The self contracted installation differs from the single source in how responsibilities are distributed between the vendor and customer. In the single source installation, all major responsibilities were given to the vendor while the customer served as an observer. In the self contracted installation, the customer retains most of the management and coordinator responsibilities and the vendor retains primary responsibility for the component installation and its integration (Fig. 12-8).

When the customer has responsibility for the management and organization of the installation, many problems transferring knowledge no longer exist. The customer will learn the characteristics of the FMS as progress is made. They will also understand the type of knowledge and expertise which is required to interact with the components.

The next section describes the chronological development of the the installation and the special precautions that were taken in reducing some of the risks associated with this installation.

12.6 INSTALLATION PROCEDURE

This procedure is a description of the FMS installation sequence which is used by a self-contracted customer. This procedure begins with detailed design and drawings and ends with a runoff for acceptance. Other installations might follow a different procedure, but the objective of this presentation is to use an actual installation as a case study to make comparisons with other available alternatives.

12.7 DETAILED DESIGN AND DRAWINGS

Once the conceptual design of the FMS was accepted, the acquisition and delivery of components was negotiated between the customer and vendors. The layout and planning group for the customer obtained floor space requirements from each vendor and constructed layouts. These were refined to coordinate with a specific location in the factory and detailed drawings were made for footings, floor height

Supplier Responsibilities

* Detailed Design
* Preparation of Site
* Component Delivery
* Component Acceptance Test
* Installation of Components
* Hardware Integration
* Communication Interface
* Component Debugging
* Computer Control Integration
* Quality Testing
* Repeatability Testing

Customer Responsibilities

* Detailed Design
* Site Selection
* Allocation of Floor Space
* Services Provision (Electricity, Heat, Offices
* Component Acquisition
* Project Management
* Release of Labor As Needed (Maintenance Operators, Engineering Managers)
* Part Programming
* System Debugging
* Quality Testing
* Run-Off for System Acceptance

FIGURE 12-8 Self-contracted type installations.

and underground and overhead service facilities. When these drawings were completed and accepted, preparation for excavation of the site began.

12.8 FOUNDATION AND FOOTINGS

The FMS customer hired a subcontractor to excavate and prepare the site for the machinery. This contractor had experience in construction sites but was not a specialist in machine footings. Figure 12-9 shows the excavation of the original floor. When this was completed, removal of dirt for trenches and footings was performed.

FIGURE 12-9 Original floor excavation.

FIGURE 12-10 Machine footings.

Footings were used for the machines and transportation system. But instead of preparing a separate footing for each station, all footings were tied together so as to appear as one large footing for the entire FMS. The footings for the row of machines are pictured in Fig. 12-10.

At the same time as the footings for the machines were being prepared, the trenches and storage units for the chip removal system were being installed.

12.9 CHIP REMOVAL SYSTEM

The FMS customer decided to use a central chip removal system. This system uses the coolant from the machines as a flush of the chips from the part into a trench below the machines. This coolant and chip mixture flow into a collection tank where the chips settle to the bottom and are then dragged out in a separation process. The coolant is then recycled for use in the machines and trenches.

The collection tank was installed at one end of the FMS with two trenches leading into it. The position of the collection tank was

FIGURE 12-11 Collection tank frame.

selected to minimize any curve and intersections needed in the trenches. Figure 12-11 shows the collection tube extending from the foundation forms. Concrete was then cast around the tube with an opening to the floor. The result is pictured in Fig. 12-12. In this photograph, the collection tube which is open at the floor can be seen extending along the line of machine footings. This tube empties its contents of coolant and chips into a central collection tank. In this tank, pictured in Fig. 12-13, the chips and oils are separated from the coolant for recycling and the chips are deposited into a waste container. This trench was left open in the machine pits but was closed in areas where no machines would be placed. When the footings, trenches and collection tank were installed, the floor was prepared. The floor contained cutouts for each machine and embedded rails, and steel supports to which machinery could be fastened. To prepare the height of these steel plates and rails, a mark was placed on one of the columns in the site. This mark

FIGURE 12-12 Concrete tank completed.

provided a single point of reference from which all measurements were taken. The transit was continually checked for proper setting to this mark to ensure that the point of reference remained exactly the same for all height measurements.

As a result of this careful checking and thorough inspection of all measurements to a single mark, the entire floor was not more than 1/4 in. out of level, and in no place was it more than 1/8 in. out of level in any six foot area. In fact, after the floor was completed, each steel plate and rail was checked for accuracy and exact shim amounts were painted on the floor to bring each point to exact height. Fig. 12-14 shows a side view of height alignment for the transport system. This allowed the supplier to prepare shims prior to the installation and to manufacture them according to precise requirements.

FIGURE 12-13 Steel liner.

FIGURE 12-14 Collection tank trench.

12.10 STATION PLACEMENT

With the completion of the floor, the construction subcontractor was finished and the delivery and installation of stations was initiated. The installation of the machines was performed by the machine tool vendor. They set the machines and constructed columns, tables and controls. Before the machines were welded to the floor, transit readings were taken from each spindle. These provided a final check on height and squareness to the plane as defined through the height of the floor. Final adjustments could then be made in shimming rather than through machine adjustments. Figure 12-15 shows the machines positioned on their respective foundations. As a result of testing machine placement, each spindle was oriented in exactly the same x, y, z coordinate plane.

The need to have each spindle operate in the same x, y, z plane is of significant importance in flexible manufacturing systems. If the installation results with stations operating in separate coordinate planes, pallets which travel to these stations must have offsets to account for these differences. X, y and z offsets can transfer

FIGURE 12-15 Machines in position.

points from one plane to points in another, but they are not sufficient for the rotation of an axis. The rotation of an axis requires an angle and can be performed through the use of complex geometry. However, this is not an easy task and can be avoided altogether by ensuring that the spindles of flexible stations operate in the same x, y, z coordinate plane. Obtaining this saved several months in debugging time and establishing offsets.

12.11 TRANSPORTER PLACEMENT

The transporter used by the FMS customer in their manufacturing system was a rack and pinion, bidirectional shuttle car. This vehicle runs on rails and obtains its commands via communication links to its controller which stays positioned at one place in the facility. Shims for the rails were calculated prior to any installation. These were set with the rails and rack. Fig. 12-16 shows the rails and rack placed on the floor and vehicle positioned to one end of the line. An AGV type vehicle with its pallet position is pictured in Fig. 12-17 and could be used as a substitute for the rack and pinion vehicle used in the current installation.

FIGURE 12-16 Rails for transporter system.

FIGURE 12-17 Transporter.

12.12 STORAGE FACILITY PLACEMENT

The work in-process storage facility provides temporary buffering of parts between a station's work table and transporter and between the transporter and work tables. The facility can be comprised of a variety of hardware, but in the FMS customers system, the storage facility consisted of a flow through pallet transfer stand. The transfer stand provides access to and from the work table and to and from the transporter. One stand is used for pallets waiting to go onto the work table and a second is used for pallets waiting for transportation. Fig. 12-18 illustrates a transfer stand used as a storage device in the FMS. In Fig. 12-19, the car is aligned with the transfer stand and a pallet is transferring onto the car. An alternative to the flow through transfer stand is a rotary pallet changer. The rotary pallet changer reduces the amount of floor space and provides a single point where the vehicle can pick-up or deliver a pallet. Figure 12-20 shows a rotary pallet changer where one side is used to store a pallet on-going to the machine and the other side is used for out-going pallets.

FIGURE 12-18 Transporter, transfer stand, pallet changer and work table.

FIGURE 12-19 Transfer Stand-Pallet Changer height alignment.

FIGURE 12-20 Transfer Stand-Pallet Changer horizontal alignment.

Work in-process was not only provided at work stations but also at general use stands. These stands made it possible to deactivate pallets during the operation of the FMS. Using the WIPAC Curve characteristics described in Chapter 3, these stands provided a means to adjust the work in-process level as adjustments to capacity occurred. The quantity and placement of these were determined through computer simulation.

12.13 LOADING AND UNLOADING AREA

The load/unload area provides the place where parts are manually removed and placed into fixtures. Each station consisted of a transfer stand and a work table. The pallet is delivered to the transfer stand where the operator pulls it onto the work table. Required activities are performed with appropriate interaction with the computer control system via terminal. Figure 12-21 shows a portion of the load/unload station with a computer terminal. The placement of a computer terminal is important to provide easy access and proximity

FIGURE 12-21 L/U station with computer terminal.

to data such as serial numbers, which need to be entered into the computer log.

12.14 HARDWARE DEBUGGING

Hardware debugging is primarily concerned with the movement of pallets throughout the facility. Hardware integration between transfer stands and work table was performed as part of the installation of the station. The next step was to integrate the transfer stand with the transporter. When this debugging begins, the major components of the FMS are in place. Figure 12-22 provides a view from one end of the line looking down the transporter with stations located along both sides.

FIGURE 12-22 Line with transporter and stations.

Along with the hardware integration was the ongoing debugging of control integration and quality testing. Described in the following sections, these tasks are not intended to be serial but rather performed parallel to each other. The fact that hardware debugging, control integration and quality testing are performed simultaneously makes the FMS installation a complex management problem.

12.15 FIRST LEVEL CONTROL INTEGRATION

First level control is the integration between the machine and its CNC or the transporter to its PLC (programmable logic controller). This integration is performed as part of the station installation or transporter installation. This is accomplished when the station hardware or transporter can achieve motion through push-button control.

In this push-button mode of operation, part programs can be run at stations and the transporter can be instructed to move, pick up or drop off pallets. This phase of debugging provides an opportunity for operators to learn how the equipment operates via manual intervention. This education should include those activities which will normally be handled by the supervisory computer and those which will remain for manual or semimanual intervention operation.

12.16 SECOND LEVEL CONTROL INTEGRATION

The second level of control integration deals with the transferring of information between the control computers. When standardized networks are used (such as MAP or Ethenet), there is less need to test protocols than when interface between computer controls is not standardized. In any case, information cannot be transferred until protocols have been fully debugged.

When it is possible to transfer information, the next step is to test the accuracy of this information by ensuring that commands can be sent to a control and the appropriate action results from the message. It is also necessary to ensure that status information can be accurately reported and recorded correctly into the receiving computer data base. When information can be transferred and its accuracy is guaranteed, the final level of computer integration can begin testing.

12.17 THIRD LEVEL CONTROL INTEGRATION

This level of control integration uses the supervisory computer to direct the activities of the facility. This means that status information can be recorded in the data base of the supervisory computer and control algorithms can recognize situations from the status data and reach conclusions. These conclusions are recorded in specific areas of the data base, from which they become communication of command to appropriate controllers.

When the first and second levels of computer integration have been fully debugged, this phase of debugging requires the control algorithm to recognize situations and provide proper responses. This debugging will be long and slow because of the need to create situations via physical conditions. With proper planning, most of this debugging can be performed through use of computer simulation. Chapter 8 describes a procedure for using computer simulation for debugging real-time control algorithms. Instead of requiring the hardware to create situations for the control algorithms to recognize and respond to, simulation can be used to create the data equivalent of these situations.

Final testing of computer integration can be done through dry run operations. These dry run operations use the entire automatic operation of the system, but no parts are loaded onto the pallets. Part programs are downloaded, but no actual operation takes place.

Instead, the station performs an operation for approximately the same amount of time the part program would take.

In this test mode, all actions are carried out by the control system, but no faults will occur within part programs. This provides an environment where operators can define their roles and responsibilities and obtain a general feel for the operation of the complex facility. This also provides the ability to test a variety of situations and observe the decision of the control algorithm. This is usually the first opportunity for the customer to see how the facility will run, and the customer will often find improvements to the control algorithms for specific situations. These suggestions should be given careful review for their likely incorporation into the control algorithm.

In the case of the FMS customer, dry runs were performed at three specific times, each lasting from six to eight hours. The first one occurred seven months after installation and the last one occurred twelve months after installation. Data was taken from these runs to indicate what production rates and which component utilization would have resulted if this were performing actual operations on real parts. This information provided the first measure of actual production capacity.

12.18 QUALITY TESTING

Quality testing starts with the unique identification of a pallet and ensures that all of its possible paths through the facility result with the same accuracy in the part. This task begins as soon as the first machine is installed and never seems to end. This task is also usually the one which extends the installation time of the flexible system.

Chapter 10 described a method for computing the number of paths for each pallet. Before the system can be used to produce parts, it must be able to exactly produce the same quality part in any of these paths.

The repeatability for the paths for a pallet requires that offsets be determined for each pallet/station combination. These offsets can account for differences between the pallets and fixtures and then reference positions to the station. These offsets can be established for x, y and z directions which will account for height, length and width differences. However, these directional offsets cannot account for rotational problems where either spindles or pallets are located in different coordinate axes at each station. The adjustments for

FIGURE 12-23 Pallet identification system.

these problems will take a very long time and are best solved by avoiding them while selling the equipment.

Usually, one master pallet is designated to bring each station to a defined level of offsets. The adjustments for each pallet are then the adjustment for the differences between a pallet to the master pallet. Because each pallet can contain a different fixture and these pallets with fixtures have unique offsets, it must have a unique identification from which each component can identify a pallet. The pallet offsets are used to reference the pallet to the machine. The fixture offsets are used to reference the fixture to the pallet. But these offsets only account for x and z directions and cannot resolve rotational alignment. When the offsets were known for each pallet and at each station, the runoff of parts was ready to begin.

12.19 RUNOFF ACCEPTANCE

The runoff of the flexible system is used to ensure that all paths are set up to produce a part within tolerance limits. Parts are loaded into fixtures and all paths are selected in a round robin method. Inspection is performed on each part. Any deviation, out-of-tolerance features and further offset calibration are detected here.

The test for productivity usually accompanies the runoff for path repeatability. However, when performance and accuracy are combined, performance becomes more important; consequently, attention is turned away from accuracy. Too much pressure is placed upon performance and often problems with accuracy remain long after the system is accepted for production.

The runoff for acceptance of the system is used for both measures of productivity and quality. Both of these are important but one cannot distract from the other. Planning and careful objectives must be established for runoff, so that the ultimate objective, the production of quality parts at desired production rates, is attained.

When the system has been accepted, management responsibility shifts from engineering to production. Now, operation becomes the subject of attention.

12.20 FMS OPERATION

The transition of management responsibility from engineering to production coincides with the acceptance runoff of the system. The acceptance is usually defined by the number of parts which meet quality specification for a designated period of time. These performance levels are based upon an efficiency level with no allowance for integration effects. In order for the system to actually achieve this level of production, it will need a component availability level of more than 90%. No highly integrated system can ever maintain such a level and, as a result, actual production during an acceptance runoff will never match targeted levels.

Because of this unique characteristic in FMS, the transition from engineering to production has no clear, decisive point. Instead, this transition occurs for a period of time until the project becomes too costly for engineering to continue to support it. There are a number of installed flexible manufacturing systems which have never achieved acceptance and final payments have never been made. Whether or not the FMS has achieved its acceptance level, eventually the production department is left with the task of managing day-to-day operations. Figure 12-24 describes these management duties in six areas: operator selection, task rotation, manual intervention,

★ Operation Selection
★ Task Rotation
★ Manual Intervention with Control System
★ Error Detection and Recovery
★ Maintenance
★ Work Incentive Program

FIGURE 12-24 Management of FMS operation.

error detection and recovery, maintenance and work incentive programs.

12.21 SELECTION OF OPERATORS

Selection of operators for integrated manufacturing systems reveals one of the myths associated with automation. Normally, the higher the degree of automation, the lower the necessary skill level of the operator. This myth has proven wrong in FMS operation.

One strategy for selecting operators for an FMS is to hire inexperienced people without habits and preconceived notions of their role in the operation. These fresh operators can be instructed in their simple and concise roles, resulting in more obedience in the operation. However, this has proved disastrous because these people cannot diagnose an unfamiliar situation. Inexperienced operators can learn and carry out their normal duties, but when something abnormal occurs, they lack the experience to diagnose the situation. In fact, some cannot even see that abnormal operations are occurring.

The fact is that integrated operation requires operators with a broad range of skills. Many of these skills will be obtained through previous experience. The dynamic operation of an FMS provides a very poor learning environment. In fact, the learning curve is so flat in an FMS that several years are needed to learn appropriate responses to certain situations.

Operators of the FMS should have broad experience in manufacturing. This includes CNC operation in a job shop, some transfer line experience, interaction with programmable logic controllers and even some interaction with computers via terminals. Certainly, there is not a pool of such operators from which to choose, so additional training is needed. But the difficult task is to make the operator who has all of these skills perform manual tasks until a situation arises that calls for his expertise.

12.22 TASK ROTATION AND TEAM CONCEPT

The FMS provides a much different operation environment than that of conventional, isolated work stations. In fact, an FMS is often treated as one large work station with many integrated components. However, to operate this integrated FMS, several operators are required for a variety of tasks.

12.22.1 Tasks in FMS Operation

The tasks in FMS operation range from general cleaning to interaction with computers via terminals. The list in Fig. 12-25 illustrates

★ Part Load/Unload of Fixtures
★ Tool Monitoring
★ Tool Replacement
★ Tool Pre-Set (depending on site)
★ Provide Information to Computer via Terminal
★ Retrieve Information via Terminal
★ Error Detection
★ Maintenance
★ Assistance with Chip Removal
★ General Cleaning
★ CNC Operation of Work Station
★ PLC Operation of Transport/Storage System
★ Inspection of Parts
★ Maintenance Station/Pallet Reference
★ Quality Reviews

FIGURE 12-25 Tasks of FMS operation.

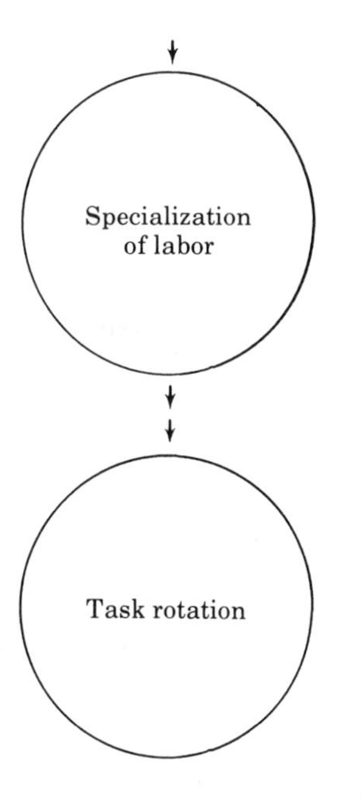

FIGURE 12-26 Assigning tasks to operators.

the wide variety of tasks required in FMS operation. Two methods for allocating these tasks are specialization of labor and task rotation (Fig. 12-26).

A specialization of labor approach assigns specific tasks to a single job. The operator with this job has sole responsibility for performing this task. For example, one job might include the task of CNC and PLC operation. Another task might include part loading and unloading, chip removal and cleaning. For each of these jobs, an operator with the necessary skill is assigned. In some cases, multiple operators might be needed if the work content is too high. Specialization of labor has the benefit of being able to assign appropriate skill levels for specific jobs. Cost of this assignment is minimized by corresponding pay rates to skill levels of a job. The disadvantage is that utilization of these operators can be high or low, depending upon their work load. In dynamic operation, this work load varies throughout the production shift and labor can become a constraining resource at some points in the FMS operation. One method to eliminate this is to use task rotation in the operation.

12.22.2 Task Rotation

Task rotation is the allocation of a group of operators to all of the tasks of FMS operation. Each operator in the FMS can perform any of the tasks required. Of course, not all operators are required to perform the same tasks simultaneously, so priorities of tasks need to be assigned. This priority changes over time so that operators rotate through the various tasks. For example, one week an operator is assigned CNC as a first priority, chip removal as a second priority and tool monitoring as a third priority. Whenever work is needed in any of these three areas, the operator must tend to it. However, if the operator is busy with a task in the second priority area and some work is required in the first priority area, efforts must be redirected to the first area.

After a period of time, customarily one week, the priorities of tasks change for each operator. This is the manner in which they rotate through the various tasks of the FMS. One week an operator might have as a first priority CNC operation, the next week part loading and unloading. This rotation of priorities creates a team approach to production.

The advantage of the team approach is that the specific job or skill level will not provide a bottleneck to the FMS production. This approach of using flexibility in the work force is consistent with the use of flexibility in the work stations. It is this characteristic of flexibility that has caused the team approach to task allocation to become a common form of FMS operation.

The disadvantage of the team approach is the cost of labor and management. The cost of labor will be higher because all operators will require the highest skill level, but might need this skill only a portion of the time that they are working. The side effect of this is worker productivity because of an operator being overqualified for some tasks. Incentives and other motivation procedures are presented in the final section of this chapter.

The second disadvantage of the team approach relates to the management, or coaching, of a team instead of individual operators. The evaluation of an individual operator's performance is difficult because his individual activity is not easily recognized. Also, workers who are members of a team can form an alliance which makes it difficult to deal with any individual without affecting the entire team. These problems are not impossible to overcome, but they do require special treatment. These treatments are commonly found in the coaching of professional athletes!

12.23 MANUAL INTERVENTION

Manual intervention is the action of the operator interacting with the computer control system. This means that components can be under computer control or manual control. Many systems permit three modes of operation: automatic, manual and semimanual. Manual and semimanual both require manual intervention and are combined for this discussion.

Computer control is a mode of operation during which the computer monitors and directs all activity within the FMS. During manual control, operators intervene to either provide information to the computer control or direct control over some activity in the system. In this intervention, the manual activity must be compatible with the computer. This compatibility is established through specific procedures for taking over a component. These procedures are referred to as manual intervention.

Manual intervention procedures contain three steps (Fig. 12-27). First, the computer control system must be informed that a component is being taken over. Secondly, the manual activity needs to be performed. Thirdly, the computer control system must be informed of status and that a component is being returned to computer control.

The specific actions of manual intervention differ for each type of component. Proper actions can only be insured through training and education. This training should include a description of the specific information which is transferred in the communication system and whether there are any situations where intervention is not permitted. One such situation might be in the downloading of a past program to a CNC. If this material is not presented as part of an education program, the operator will assume that the work station can be operated as an isolated device. Previous experience might lead an operator to decide to take over a function. The operator is skilled at the activity needed for manual control, but might not appreciate the need for notifying the control system of such action.

This potential problem of an operator not cooperating with the control system is overstated, and the reason many FMS owners elect to use inexperienced operators who do not have any bad habits. This, however, is an overreaction to a potential problem.

Manual intervention procedures must have all these steps performed. The experienced operator can perform the manual activity. The operator must be educated as to the procedures for notifying the control system of takeover and return to computer control, but this can be handled through proper training which emphasizes the role of the operator. The inexperienced operator will always remember to notify the supervisory computer but will not know what to do for the actual manual control. The effectiveness of the inexperienced operator for manual intervention will be much less than that of the experienced operator.

The role of operators is an important issue to address in the training and education of FMS operators. It is easier to address a

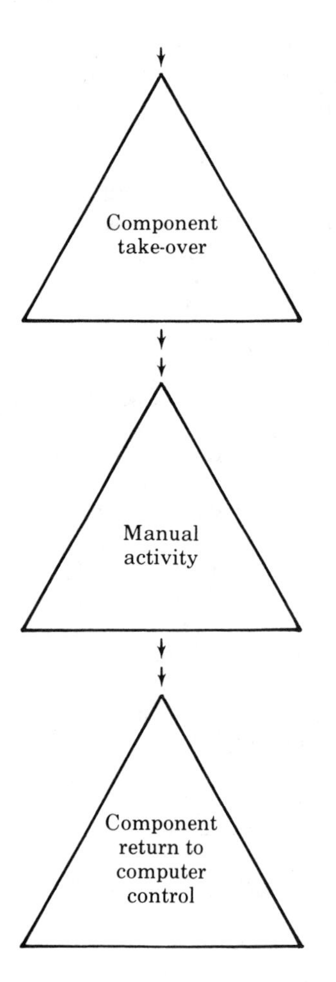

FIGURE 12-27 Steps for manual intervention.

new role than to educate an inexperienced operator in manual activity. Education of operators is one side of the manual intervention issue.

The other side of manual intervention is in the computer control. The computer control must be designed to permit manual control over one component while maintaining computer control over the others. Even when the component is under manual control, the control system should remain in communication with the component. This enables status information to be collected and retained during periods of manual intervention. Many control systems provide for either total control or no control. This is simplest for the software, but there is no real, practical reason for this approach. The computer control should retain constant communication with each component and in this way monitor manual intervention activity.

The control system must recognize that computer control is appropriate for some periods of operation. But there is a real need to provide a combination of computer control for some components and manual control for others. The software must be specified to facilitate this type of operation, and the intervention procedures must be defined and transferred to operators through education and training. Avoiding this issue will only delay the time when it will eventually need to be addressed.

12.24 ERROR DETECTION AND RECOVERY

Error detection is the process of identifying when computer control of a component is no longer possible. Recovery is the process of returning the component to computer control via manual intervention. Often the most difficult part of this problem is the detection of an error (Fig. 12-28).

Error detection is most visible when a component fails to perform its required action. In order to observe the error, the observer must notice a need for a component to function and then observe that this expected action is not taking place. This is not easy and requires either constant attention to system operation or extensive experience in recognizing situations during a short period of observation.

Assistance is available for error detection. Most stations will provide a beacon signal which can either be sounded or lit when the control system detects an error. The detection of these errors is initiated from a local controller unable to carry out some task. Usually this detection is based upon an elapsed time or time out of the anticipated completion of an activity.

The detection of these time-out errors is easily noticed by operators. However, this type of error detection is based upon the foresight of the control programmer. The most difficult errors to detect are those which occur when there is no planned monitoring of a task. In these cases, error detection relies only on the experience and skill of the operator. For example, the protocol of communication between a supervisory computer and a CNC can become mixed up. The supervisory computer might be waiting for an end of program message to verify that the past program has been received, but the CNC might be expecting more information from the supervisory computer. When this situation exists, the CNC might appear to be under automatic control, but no activity will take place

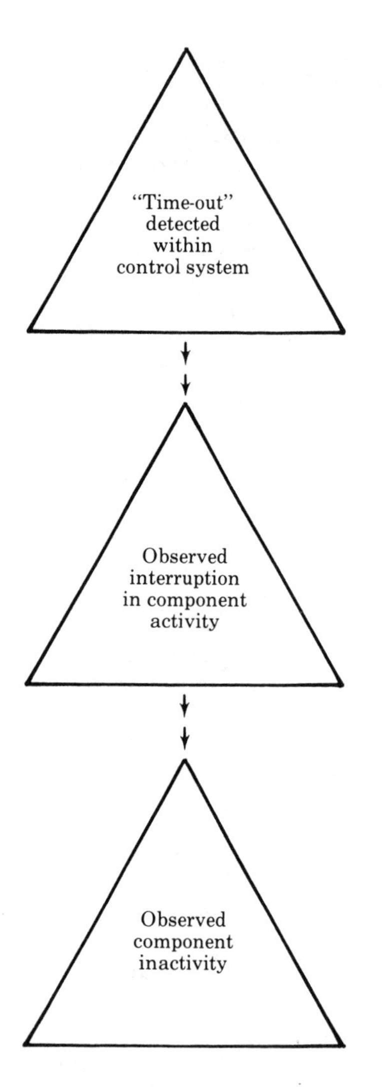

FIGURE 12-28 Methods for error detection.

- ★ Take Control of Components
- ★ Get Component to a "Known State"
- ★ Diagnose Status
- ★ Determine Mechanical or Control Related Problem
- ★ Signal Appropriate Assistance
- ★ Return Component to Computer Control

FIGURE 12-29 Recovery procedure.

at the station. This error is detected through the observation that something should be happening. The station might be held in this suspended situation for several hours before an operator notices the error.

Certainly, some enhancements to the control system could be implemented to detect this error. But there will always be errors which the control system cannot detect. For these errors, the operator's skil and experience determines the methods and timelines of detection.

Once the error has been detected, the next phase is recovery (Fig. 12-29). The recovery process starts with an attempt to get the component with the error to a known state, usually accomplished through manual control. In other words, the operator takes control of the component and performs some manual function. This determines whether the problem is mechanical or control related. Again, with the experience and skill of the operator, the problem can be further defined.

Once the problem has been resolved, the last step in recovery is to return the component to computer control. However, not all problems can be resolved by the intervention of an operator. Some require the assistance of maintenance personnel. The role of maintenance in the FMS is the subject of the next section.

12.25 MAINTENANCE

Maintenance is the process of preserving machinery, repairing mechanical components and making enhancements to the software control (Fig. 12-30). Many maintenance plans do not include software as part of this area, but the software will have problems or

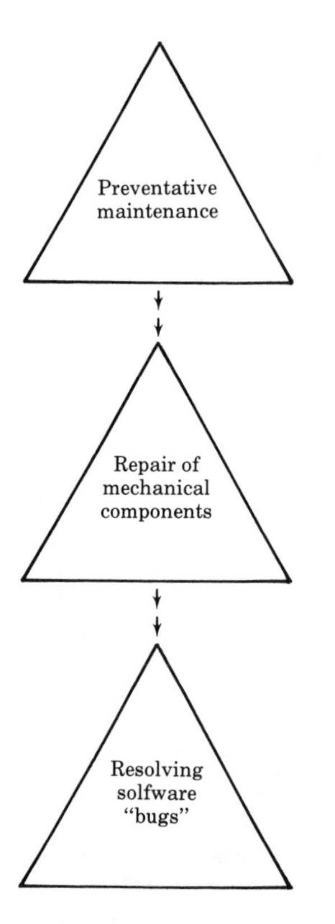

FIGURE 12-30 Maintenance tasks.

bugs just like the hardware. Therefore, these must be dealt with in an ongoing process.

Preventative maintenance involves the routine activity of preserving the operation of components. Because these tasks are routine, they are often shared between the operators and the maintenance personnel. The specific tasks which are shared depends upon the job assignment of the operators and the role which the maintenance department assumes.

The repair of mechanical components is the most common function of maintenance. This repair involves specialists in such areas as electrical and mechanical procedures and pipe fitting. In the case of FMS operators, some maintenance personnel in each of these areas might be specifically assigned. Whether the maintenance personnel is dedicated to FMS depends upon management's perspective of machine availability and its relation to production. When management perceives that production is highly dependent upon machine availability, then maintenance personnel will be dedicated to the FMS. But as management realizes that some integration effects are going to occur in the FMS operation and the increase of the response time for maintenance has little impact upon production, they will treat maintenance as a service to the operator, rather than as an expense.

The third area of maintenance is with the software control. This type of maintenance is usually provided by the software developer rather than through internal services. For this reason, software is not handled the same way as hardware in the detection of problems.

When the maintenance personnel are not on site, the response time needed to recover from software problems is typically longer than that for hardware. Therefore, attempts are made to bring the software maintenance tasks to within the internal operation of the FMS. This is done through the acquisition of the source code and programmers with skill in real-time control. However, the acquisition of this service is expensive. The decision to acquire this type of maintenance is again based upon the perception of response time's impact upon production and the direct cost of the service.

12.26 WORK INCENTIVE PLAN

Providing the resources and maintenance support for the FMS operation is one matter, and obtaining the efficient use of these services is another altogether. How can operators be motivated to learn and provide timely activities to insure efficient FMS operation? The answer to this question is in the work incentive plan.

Use of work incentive and FMS operation are perceived as incompatible with one another. The reason for this perception comes from the basic principle of work incentive plans: the need to relate an amount of pay for a single part. Because of the nature of FMS, no single operator can be assigned a part and no single part can be assigned to an operator. For this reason, traditional incentive plans based upon piece and part production have not been effective in the FMS. An innovative method of incentive is needed which addresses the unique operation environment of the FMS.

The Manufacturing Integration Model (MIM) states that production is a function of gross capacity, station availability and integration effects. Integration effects are further defined as derived from inventory, balanced loads and flexiblity, and these effects are quantified with inventory and flow time. By combining these results, production is a function of gross capacity, station availability, inventory and flow time.

The fact that four variables determine production is one reason why conventional incentive plans are not effective in the FMS. The reason production is a function of these four variables instead of the customary three (given capacity, station availability and inventory) is due to flexible integration.

Flexible manufacturing systems use flexible integration as a means to reduce work in-process inventory levels. Integration has been used in the transfer line to reduce inventory levels but requires the use of balanced operation times to eliminate integration effects. The use of flexiblity eliminates the need for balanced operations by providing alternative paths in the production process. However, the presence of alternative paths will not eliminate integration effects.

A work incentive plan must provide an incentive to increase productivity. In the case of FMS, production is derived from four variables. Each of them must be reviewed as to its suitability to a work incentive plan, shown in Fig. 12-31.

The first variable from MIM is gross capacity. Gross capacity is the amount of production which can theoretically be obtained from the production system. Production can be increased by increasing gross capacity, but this is an area where operators have

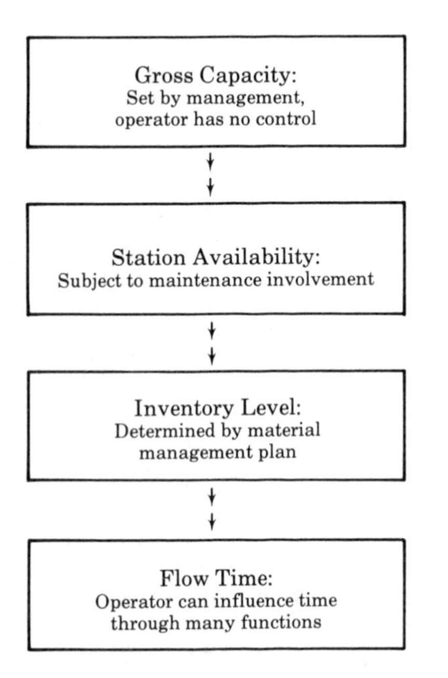

FIGURE 12-31 Work incentive plan with MIM.

no control. Gross capacity can be increased through shorter operation times or acquisition of additional equipment, but these are engineering changes, not operational changes.

The second variable from MIM is station availability. Station availability is the amount of production which can theoretically be obtained when station failures occur. In order to increase production, station availability must increase or frequency of failure and repair time must decrease. The operators might have some control over frequency of failures if they are involved in a preventive maintenance plan, but this is only a small portion of the station availability problem. Maintenance personnel would have to be involved in an incentive plan based upon station availability and operators would not have individual incentive for production.

The third variable of MIM is inventory. The higher the inventory, the lower the integration effects, resulting with increases in production. Any incentive plan based upon inventory will defeat the overall objective of flexible integration, which is to reduce inventory through flexibility.

The fourth variable in MIM is flow time. Flow time is the elapsed time from part introduction to the system until part completion. A decrease in flow time increases production. Flow time is comprised of station time plus storage time plus transport time. These are the traditional areas where operators have incentive to reduce operation times. The operator can influence station time by responding to faults, maintaining tooling to prevent took breakage, and providing efficient operation during periods of manual intervention. The operator can influence storage time by maintaining a constant velocity of the parts moving through the system. By reducing delays at stations, they will indirectly reduce the integration or carry-over effects, which in turn will reduce the storage time of the parts. The transportation time is the total time required to transport a part between all of its operations. The operator will not have much influence over this time. Flow time is an observable outcome of the FMS operation and is directly related to production. Operators' tasks and activities can have direct influence over this time. This seems to be a potential basis for a work incentive plan; however, some further issues must be addressed.

12.26.1 Opposing Arguments

One opposing argument to this plan is that operators will run only one pallet in the FMS to obtain a minimum flow time. Incentive pay is based upon a single part and if operators can make the most money with only a single pallet, then this means that they are maximizing production as well. The WIPAC Curve, described in Chapter 3, shows the highest net production for inventory and flow time levels. Given that the inventory level stays fixed and flow time decreases, production will increase. If flow time decreases to the point that the inventory level is not sufficient to consume all available capacity, operators will recognize that they can make more money with an increase in inventory. So the argument that operators will reduce inventory below a level needed to consume capacity is not a likely situation.

Another argument against a work incentive plan is that operators might intervene too frequently and take over routing and other computer controlled activities. Operators are likely to take on more tasks only when they know the return (perceived pay) is greater than the effort. If it is possible for an operator to perform a task manually and more efficiently than a computer control, it is in everyone's best interests to let the operator do it. An operator should not be constrained by the inefficient operation of computer control.

A third argument is the production of inferior quality parts. However, this argument can be made against any work incentive plan. The response is simply that the part must meet quality guidelines before being counted for incentive pay.

The use of rework in the FMS can be part of the incentive plan as well. No incentive was given for the rework part during its first time through the system, but it should be possible to obtain incentive pay for its rework process. An assessment of the amount of rework needed and an estimated flow time must be established for the reworked part. Incentive pay is provided when the actual flow time is less than the estimated time.

12.26.2 Implementation of the Work Incentive Plan

Implementation of such an incentive plan requires thorough evaluation (Fig. 12-32). First, the incentive rates must be established from the WIPAC Curve and computer simulation. Given a level of station availability, the WIPAC Curve can be used to determine the appropriate inventory and flow time levels. These figures will allow for some integration effects and provide a realistic level of production. This flow time should then be used as the base rate. If flow time decreases by 10%, production must increase by 10% for a given inventory level. The value of the 10% increase in production can be computed and shared with the operators. This amount is then the incentive pay for any part which obtains a flow time 10% less than the base. Figure 12-33 shows the incentive pay chart based upon a variety of flow times.

When engineering changes are made that affect flow time, the new WIPAC Curve is needed to re-establish incentive pay rates. The same procedure used in establishing the initial pay rates needs to be performed with the engineering changes included.

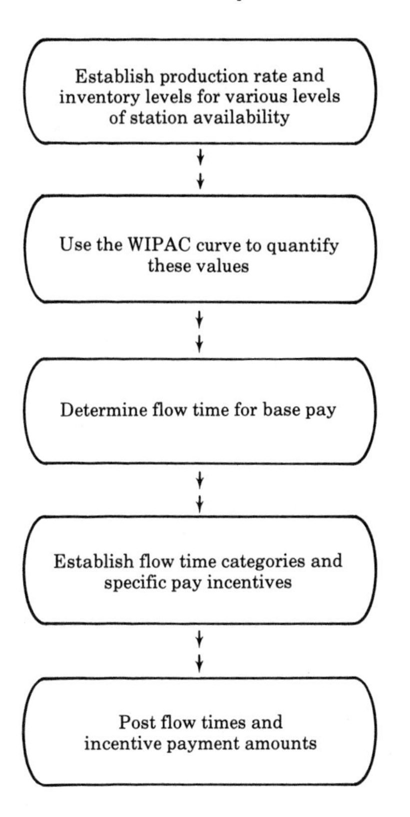

FIGURE 12-32 Implementation of the work incentive plan.

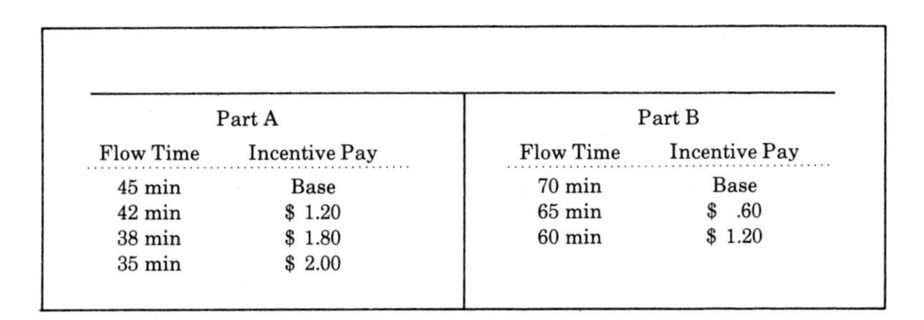

Part A		Part B	
Flow Time	Incentive Pay	Flow Time	Incentive Pay
45 min	Base	70 min	Base
42 min	$ 1.20	65 min	$ .60
38 min	$ 1.80	60 min	$ 1.20
35 min	$ 2.00		

FIGURE 12-33 Work incentive plan.

A work incentive plan based upon flow time is a proposed solution to a complicated problem. Only the basics are presented, and many details for implementation need to be customized for specific application. The purpose of this presentation is to provide insight into a plan that applies incentive to an independent variable of the dependent production variable.

12.27 CONCLUSION

The installation of an FMS is a complicated and time-consuming task, which contributes to its high risk. Phased in installation may reduce some risk, but the resulting practical interruption of the operation is much too costly to offset the risk reduction. Because of the high degree of integration, FMS capacity has been acquired in a single phase with a long installation period.

The installation itself starts with the excavation and construction of the foundation of the system. Much planning is needed in this step because of the amount of underfloor service necessary in the FMS. One of the most important tasks in the placement of equipment is its proper orientation to the other components.

When individual spindles are used, these must be aligned in the same x, y, z coordinate plane. If they are not, then a long time is needed to calculate offsets for pallets, fixtures and tools. However, each of these offsets cannot make adjustments for a rotation of an axis. With proper measurements, the height alignment can be obtained prior to any equipment installation, and adjustments and foundation imperfections can be known prior to equipment delivery.

The integration of the system is performed through three steps. The first step is integration of the component with its own controller. The second step is the establishment of a communication network between the controller and supervisory computer. The third step is use of the supervisory computer to receive status information, make decisions and communicate the decisions to respective components. The use of dry run testing is good for debugging and third level of control integration because random failts are eliminated in the operation.

Runoff and quality acceptance are the tasks which initiate the transfer from engineering to production. These must include allowances for station availability and integration effects which will occur during the runoff period. They cannot be based simply on a parts per hour measurement.

After runoff and system acceptance, the responsibility of the FMS moves from engineering to production management and operational control. This involves operator selection, tasks and activity description and operator education. Many flexible manufacturing systems have used a team concept, where operators rotate through a variety of responsibilities. This has provided some solutions to situations where operators have greater skills than those which are needed for some tasks. A better solution is to obtain specialized labor instead of using an operator with skills beyond those needed for assigned tasks.

Effective procedures are needed for error detection and recovery. Because of the high degree of integration and many levels of computer control, not all problems can be sensed. The operators must be able to recognize error situations and have sufficient skill in identifying the specific cause of the error.

Finally, the FMS can provide an incentive pay environment where operators are paid based upon their efforts. This plan is based upon the flow time of parts through the FMS. As the flow time decreases for a given inventory level, production will increase. Operators can reduce flow time through quick responses to faults, tool maintenance and informed operation. The revenues from increases in production can be shared with the operators, and through computer simulation, specific pay incentive can be determined for various flow times.

Overall, the FMS operation is complicated by the presence of integration effects. However, once these are understood and quantified, they simply become adjustments to the manufacturing system.

Glossary

Aggregate planning: Forecasted production's demand upon a set of stations using time standards.

Artificial intelligence: Emulation of deductive reasoning by use of computer programs.

Automation: Implementation of a control system to coordinate mechanical action with use of sensor feedback.

Balanced loads: The uniformity between aggregate planning amounts for individual stations.

Blocked station: A station has completed its operation, but can not clear its work table due to material handling delay or saturated storage facility.

Color Graphic Animation: Bringing of graphical images over a background layout according to simulated activity.

Computer Numerical Control (CNC): Computer designed for programmable control of machine tool activity. Resolver/stepping motors used as sensor devices.

Control Algorithms: A specific procedure (usually implemented in software) to make a decision from a current situation.

Degree of Automation: The amount of activity which is monitored and directed by sensors with a computer system.

Download: Transfer of a part program for a central computer to a station controller such as a CNC.

Efficiency factor: Percent of time a factory component (station, operators, transporter) will be available during a production period.

Fixtures: Clamping mechanism for securing a part in a specific orientation.

Fixture offsets: X, Y, Z adjustments from a reference point to account for individual characteristics of a fixture.

Flexibility: Alternative means to complete a required operation.

Flexible manufacturing: Manufacturing strategy which utilizes flexibility as a substitute for inventory.

Flexible Manufacturing System (FMS): Implementation of Flexible Manufacturing by use of computer controlled hierarchy, integrates a material handling system with flexible stations where the control system monitors activity, and directs movement of parts.

Gross production: The maximum production rate where the bottleneck station is utilized 100% of the total time.

Integrated manufacturing: Use to describe a production facility where the activities at one station directly impact the activities at another.

Integration effects: Adjustments to production capacity due to carryover effects of one station's activity into another.

Inventory: The number of parts which are available for an operation.

Island of automation: Focused area of a production facility which uses automation to integrate the activity of a set of station.

Just-In-Time (JIT) Techniques: Shop floor techniques which reduce inventory levels with use of simple material tracking and monitoring procedures.

Ladder logic: Programming language use for programmable controller. Use for emulation of logical relay panels.

Manufacturing Integration Model (MIM): A conceptual view of manufacturing which accounts for integration effects and provides a means to plan for these effects.

Material shortage: Loss of inventory which causes a station to become idle.

Material Requirement Planning (MRP): Procedure for determining material needs during a production period. These needs are used to prioritize parts in the process.

Mechanization: Replacement of human activity with machine. Does not require a feedback active system.

Net production: The observable production rate resulting from the actual operation of the facility.

Optimized Production Technique: Shop floor technique which optimizes the production of the "bottleneck" station with use of computerized part tracking system.

Pallet offsets: X, Y, Z adjustments from a reference point to account for the individual characteristics of each pallet.

Pallet: A device to provide uniform material handling of parts. Can also provide a uniform surface for fixtures.

Part program: Computer program which directs the CNC to carry out a series of outlines. Languages include APT, Compact, and is commonly used for metal removal actions.

Planning horizon: The length of time the production facility has to produce a planned number of parts. This time is adjusted by the efficiency factor to provide a net planning period.

Pots: Position in a tool changer.

Programmable controller: Computer device to carry out the directions of ladder logic programming. Replaces the need of a relay panel.

Sister tools: Identical tools in the same tool changer which can be used to substitute for each other.

Situation decisions: Response to a particular status of parts, stations, and transporters.

Station unavailability: Adjustment to gross production due to station failure, tool and operator availability, scrap parts, and rework.

Storage factor: Ratio of time a part spends in storage to the time spent in operation while in the production facility.

Time standards: Allowed time for an operation to be performed on a part.

Tool changer capacity: Number of positions which are available. An individual tool can occupy one, or two, or three positions due to its diameter.

Tool life: The amount of accumulated use which a tool can obtain before it must be serviced.

Tool offsets: The adjustment from a reference point to account for lengths of individual tools.

Tool window: Time duration for forecasting tool requirements from a planned production schedule.

Transfer line: Integrated manufacturing system where stations are located along a line and parts require a sequence of operation which is consistent with the placement of stations.

Upload: Transfer a part program from a station controller such as CNC to a central computer used to distribute a part program to another station.

Window of inventory: Interruption of uniform dispersion of inventory throughout a facility. Its occurrence causes a station to become idle due to a material shortage.

Bibliography

Achatz, R. and D. J. Parrish (1987): Host Computer Comtrols FMS at All Levels, FMS Mag., 5:21-25.

AGVS-T (1984): Modern Materials Handling, *39*:55-61.

Albus, J. S. and M. L. Fitzgerald (1982): An Architecture for Real-Time Sensory-Interactive Control of Robots, IFAC Information Control Problems in Manuf. Tech., pp. 81-86.

Bergstrom, Robin P. (1984): FMS: Taking Stock of FMS...Users Speak Up, Manuf. Engineering, pp. 48-55.

Bevans, John Patrick (1982): First, Choose an FMS Simulator, American Machinist, pp. 143-145.

Birge, John and Richard C. Wilson (1981): Minutes on Workshop on Modelling and Simulation of Manufacturing Systems, University of Michigan, Ann Arbor, Michigan.

Blumberg, Melvin and Donald Gerwin (1982): Coping with Advanced Manufacturing Technology, School of Business Administration, University of Wisconsin, Milwaukee, Wisconsin.

Bonetto, R., et al. (1984): Simulation of an FMS: Case Study of the Citroen Factory, Advances in Production Management Systems.

Burgam, Patrick (1984): FMS Control: Covering All the Angles, CAD/CAM Tech., pp. 11-14.

Burstein, Michael (1986): Guidelines for Estimation of Costs in the Economic Justification..., FMS Conference Presentation, Rosemont, Illinois.

Buzzacott, J. A. (1982): "Optimal" Operating Rules for Automated Manufacturing Systems, IEEE Transactions on Automatic Control, *AC-27*:80-86.

Carrie, A. S. and E. Adhami (1983): Simulating an FMS, IFS Conference Presentation, Bedford, England.

Chan, W. W. and K. Rathmill: Digital Simulation of a Proposed Flexible Manufacturing System, Department of Mechanical Engineering, UMIST, Manchester, England.

Cross, Kelvin F. (1984): Production Modules Offer Flexibility, Low WIP for High Tech Manufacturing, IE, pp. 64-72.

Davis, G. I. and J. R. Buck (1980): Man-Machine Simulations in Industrial Systems, volume 1, Purdue University, Indiana.

Demonet, Gene (1986): Flexible Manufacturing Systems for Super-Alloy Jet Engine Parts, SME/FMS Conference Presentation, Rosemont, Illinois.

Diebold, John (1952): Automation, The Advent of the Automatic Factory, Van Nostrand, New York.

Diesch, K. H. and E. M. Malstrom (1984): Physical Modeling of Flexible Manufacturing Systems, IIE Fall Conference Presentation, Iowa State University, Ames, Iowa.

Dubois, Didier (1983): Mathematical Model of an FMS with Limited In-Process Inventory, European Journal of Operational Research, *14*: 66-78.

Dunn, Peter (1985): FMS Simulation-Yet Another System, FMS Mag., *3*:198-200.

Dupont-Gatelmand, C. (1982): A Real Time Control System for an Unaligned FMS, IFAC Information Control Problems in Manuf. Tech., Maryland.

Dyke, R. M. (1967): Numerical Control, Prentice Hall, New Jersey.

Erschler, J., et al. (1984): Periodic Loading of Flexible Manufacturing Systems, Advances in Production Management Systems, pp. 401-413.

Ettlie, John E. (1985): Organizational Adaptations for Radical Process Innovation, presented at National Meeting of Academy of Mgmt., San Diego, California.

Ettlie, John E. (1985): The Implementation of Programmable Manufacturing Innovations, Industrial Technology Institute, Ann Arbor, Michigan.

Ettlie, John E. and Janet L. Eder (1985): Managing the Vendor-User Team for Successful Implementation of..., AIM Tech 22 Conference Presentation, St. Louis, Missouri.

Finkel, James I. (1983): Computer Models for Flexible Manufacturing, CAE, pp. 52-60.

Flexible Approach to FMS Design (1982): Machinery and Production Engineering, pp. 166-167.

Gaines, B. R. and M. Shaw (1986): Knowledge Engineering for an FMS Advisory System, Simulation in Manufacturing Int'l Conference, Chicago, Illinois.

Gershwin, Stanley B. (1982): Material and Information Flow in an Advanced Automated Manufacturing System, Massachusetts Institute of Technology, Cambridge, Massachusetts.

Gerwin, Donald (1984): A Theory of Innovation Processes for Computerized Manufacturing Technology, School of Business Administration, University of Wisconsin-Milwaukee, Milwaukee, Wisconsin.

Glenney, Neil (1981): Modular Integrated Materials Handling System Facilitates Automation, IE, pp. 118-119.

Grant, H. (1986): Production Scheduling Using Simulation Technology, Simulation in Manufacturing Int'l Conference, Chicago, Illinois.

Hayes, Robert H. (1984): Manufacturing Processes-Technological Innovations, Restoring Our Competitive Advantage.

How to Justify Installing FMS (1982): Production Engineer, *61*:30-32.

How to Realize the Benefits of FMS (1983): Production Engineer, *62*: 44-45.

How the Unions View FMS (1983): FMS Mag., pp. 144.

Hughes, John(1987): It's Methodology, Not Technology, That Counts, FMS Mag., 5:81-85.

Hutchinson, G. K. (1978): Proceedings from FMS Workshop, FMS Workshop Presentation, Peoria, Illinois.

Hutchinson, G. K. and A. T. Clementson (1984): Manufacturing Control Systems: An Approach to Reducing Software Costs, International Conference Presentation, MIT Cambridge, Massachusetts.

Hutchinson, G. K. and John R. Holland: The Economic Value of Flexible Automation, Manuf. Sys. *1*:2.

Kimemia, Joseph and Stanley B. Gershwin (1983): Algorithm for the Computer Control Flexible Manufacturing System, IIE Transactions, *15*:353-362.

Klahorst, H. Thomas (1981): FMS: Combining Elements to Lower Costs, Add Flexibility, IE, pp. 112-115.

Lawrence, J. and J. R. Dawson (1982): The Interrelationship Between Components in a FMS, Engineer's Digest, Great Britain, *43*:9-11.

Lenz, John (1986): General Theories of Flexible Integration, Jacobs Engineering Group, Inc., Pasadena, California.

Lenz, John (1987): Automatic Evaluation of Simulation Output or "The Answer to Why", simulation in Manufacturing International Conference, Turin, Italy, pp. 3-11.

Lenz, John (1983): FMS Design Using Microcomputer Graphics, AUTOFACT 5 Conference Presentation, Detroit, Michigan.

Lenz, John (1986): Benefits of FMS: Reducing Work-In-Process Levels, FMS Mag., *4*:83-86.

Lenz, John (1983): MAST: A Simulation Tool for Designing Computerized Factories, Simulation Mag., *40*:51-58.

Lenz, John (1983): Information Needs of FMS Operators, CAM-1 Conference Presentation, St. Louis, Missouri.

Lenz, John: Using Animated Simulation to Model FMS Designs, Automated Materials Handling and Storage, Pennsauken, New Jersey.

Lenz, John (1981): MAST: A Simulation Language, ORSA Conference Presentation, Toronto, Canada.

Lenz, John (1982): Design and Control of Manufacturing Systems Using Simulation, SME Conference Presentation, Los Angeles, California.

Mohri, N., et al. (1982): Design and Control of Manufacturing Systems Using Simulation, SME Conference Presentation, Los Angeles, California.

Mohri, N., et al. (1982): Inprocess Monitoring of Tool Breakage Based on Auto-Regressive Model, IFAC Information Control Problems in Manuf. Tech., Maryland, pp. 41-45.

Mollo, Marco, et al. (1983): Modular Software for FMS Applications, FMS Mag., pp. 217-221.

Montag, Alfred C. (1984): Flexible Automation for High-Volume Production, Manuf. Engineering, pp. 79-80.

Morgan, T. K. (1984): Planning for the Introduction of FMS, FMS Mag., pp. 13-15.

Mortimer, John (1984): Building a Framework for FMS in High Volume Industries, FMS Mag., pp. 28-30.

Mosier, Charles T. and Brian Hoyler (1986): Machine Cell Formulation for FMS Design, FMS Conference Presentation, Rosemont, Illinois.

Musselman, Kenneth J. (1984): Computer Simulation: A Design Tool for FMS, Manuf. Engineering, pp. 117-120.

Noble, David F. (1984): Forces of Production, Alfred A. Knopf, new York.

Noble, I. A. (1984): Flexible Manufacturing Systems in the Foundry Industry, British Foundryman, 77:137-141.

Olig, Gene A. (1984): An Overview of an FMS System, The Carbide and Tool Journal, Giddings and Lewis Co., Wisconsin.

Olling, Gustav (1978): Experts Look Ahead to Day of the Full-Blown Computer Integrated..., IEEE Spectrum, *15*:60-66.

Panov, A. A. (1983): Production Improvements Through the Intro. of FMS and Industrial Robots, Soviet Engineering Research, *3*:50-52.

Perotto, Giovanni and Adriano DeLuca: FMS Simulation Advanced Mathematical Models and Fast Simulation..., EICAS Automazions S.p.A. Presentation, Torino, Italy.

Production Costs are Halved by FMS (1983): Machinery and Production Engineering, *141*:29-32.

Ross, Edward A. (1986): The FMS Market Opportunity: A Well-Kept Secret, FMS Conference Presentation, Rosemont, Illinois.

Salomon, Daniel and John E. Biegel (1984): Assessing Economic Attractiveness of FMS Applications in Small..., IE, pp. 88-96.

Scott, Harold A., et al. (1983): Hierarchal Control Model for Automated Manufacturing Systems, Computer and Industrial Engineering, 7:241-255.

Seliger, G. et al. (1986): Decision Support for Planning Flexible Manufacturing Systems, Simulation in Manufacturing International Conference, Chicago, Illinois.

Shannon, Robert E. and Richard Mayer (1985): Expert Systems and Simulation, Simulation, pp. 275-284.

Simpson, J. A. (1982): National Bureau of Standards' Automation Research Program, IFAC Information Control Problems in Manuf. Tech, Maryland.

Simulation in Manufacturing (1987): Royal Swedish Academy of Engineering Sciences, Stockholm, Sweden.

Sloggy, John E.: How to Justify the Cost of an FMS, Giddings and

Lewis Company, Janesville, Wisconsin.

Thon, Heinz-Jurgen and Reinhard Willinger (1983): FMS-Modular Automation for the Factory of the Future, Siemens Power Engineering, 5: 330-331.

Thornton, Jack (1985): Simulation and FMS Planning...Arizona State Uses PC, CIM, pp. 8-9.

Vasiliev, V. N. (1985): Implementation of FMS in Soviet Industry, FMS Mag. 3:101-103.

Warnecke, H. J. and H. Kampa (1982): Flexible Production Systems in the Federal Republic of Germany, IFAC Information Problems in Manuf. Technology, Maryland.

Wichmann, K. E. (1987): An Intelligent Simulation Environment for Design and Operation of FMS, SCS International Conference Presentation, Vienna, Austria.

Wichmann, Knud Erik and John Lenz (1987): Integration of Fortran Based Simulation Program and Prolog Based XMAS, Working Paper in submission.

Wilhelm, Wilbert and Subhash C. Sarin (1983): Models for the Design of Flexible Manufacturing Systems, AIIE Conference Presentation, Ohio State University, Columbus, Ohio.

Young, Robert E. and S. F. Simon Chen (1984): A Supervisory Control System for a Flexible Manufacturing System, AIIE Conference Presentation, Chicago, Illinois.

Index